T0089328

ANATOMY 101

FROM **MUSCLES AND BONES** TO
ORGANS AND SYSTEMS, YOUR GUIDE
TO **HOW THE HUMAN BODY WORKS**

KEVIN LANGFORD, PhD

Adams Media
New York London Toronto Sydney New Delhi

▲adamsmedia

Adams Media
An Imprint of Simon & Schuster, Inc.
100 Technology Center Drive
Stoughton, MA 02072

For information about special discounts for bulk purchases, please contact Simon & Schuster Special Sales at 1-866-506-1949 or business@simonandschuster.com.

The Simon & Schuster Speakers Bureau can bring authors to your live event. For more information or to book an event contact the Simon & Schuster Speakers Bureau at 1-866-248-3049 or visit our website at www.simonspeakers.com.

Interior photos © iStockphoto.com and 123RF

Manufactured in China

20 19 18 17 16 15 14 13 12

Library of Congress Cataloging-in-Publication Data has been applied for.

ISBN 978-1-4405-8426-8
ISBN 978-1-4405-8427-5 (ebook)

Contains material adapted from *The Everything® Guide to Anatomy and Physiology* by Kevin Langford, PhD, copyright © 2015 by Simon & Schuster, Inc., ISBN: 978-1-4405-8182-3.

CONTENTS

INTRODUCTION

THE BUILDING BLOCKS OF ANATOMY AND PHYSIOLOGY

The human body has always amazed mankind. Early scientific drawings and diagrams demonstrate the long-standing fascination with the body. Even cave drawings and later hieroglyphs illustrate that people were aware of the complex machinery of the human body. Our fascination continues to the present day, as we dig ever deeper into learning everything we can about the human body. Our understanding has advanced dramatically in just the last 20 years alone.

The study of the human body is divided into two different but closely related disciplines. Human anatomy is the study of the structure of the human body while physiology is the study of its function. Together, they help us understand how the human body works. In this book, you won't just learn the structure of the human body and the functions of its various parts, you'll also discover *why* it does what it does.

Cells, tissues, and organs are often intricately arranged to facilitate many functions simultaneously; complex biochemical processes take place that enable your body to perform those functions. In *Anatomy 101*, all of these processes and structures of the human body are explained. After reading this book you'll know the human body inside and out.

The amount of complexity can seem overwhelming when you're studying anatomy and physiology, especially at first, and particularly if you don't have a strong background in biology. Don't be intimidated! This book is designed for the reader who doesn't already have a PhD in biochemistry. Even if it's been a few decades since high-school biology, with careful reading, you'll be able to grasp the principles described in this book. By starting with a solid foundation, you will eventually master the intricacies of the human body. Don't forget that you already have a head start: you own a human body.

While it may seem obvious that the human body is made of organs and the structures that connect them to each other, this book doesn't start there, at the macro, big-picture level. It starts at the micro level, inside your very cells, with a description of the processes at work that help your body's cells know what to do, when, and how. We'll look at the biochemical basis of human life—the organic and inorganic elements, compounds, and molecules that are necessary for the functioning of your body. We'll look at how cells communicate and replicate. That will help create the solid foundation you'll need to understand the rest of the material.

Once those building blocks are in place, we'll move on to a discussion of tissue, the foundation of all the organs in your body. Once this material is covered, we'll move on to the major systems in your body, including the skeletal, nervous, cardiovascular, and respiratory systems (among others).

For each system, common diseases and disorders are also described. Related material, such as how the senses integrate into the sensory system and the importance of nutrition to human health, are also covered.

Consider this book your one-stop information source for understanding the human body from cranium (head) to phalange (toe).

THE CHEMISTRY OF CELLS

Nuclear Reactions and Why We Love Them

Everything in the universe—from the largest of stars in the sky to the smallest grain of sand on the beach—is made up of matter. To be more precise, everything that takes up space and has mass is made up of matter. That small grain of sand may not seem like it takes up any space or has any real mass, but wait until it gets in your shoe. Then you'll know it as a physical object.

We might call matter "physical substance" (as opposed to that random thought you just had about what's for lunch; random thoughts have no physical essence).

The study of the structure of matter, its properties, and how different kinds of matter interact is called chemistry, and an understanding of basic chemistry is crucial to learning the principles of anatomy and physiology.

The interaction of atoms—which you probably know as the building blocks of matter—has created the human body and the world it inhabits. Atoms together form elements, a type of matter that cannot be broken down by chemical means (that's where the nuclear reaction comes in—elements can only be changed by nuclear means). Various elements combine together to create cells, which are the smallest structural units in the human body that perform a function. For example, your blood cells carry oxygen throughout your body. They have a distinct structure from cells that perform other functions, such as nerve cells or muscle cells. Chemistry rules not only how these cells are structured but also how they perform their functions.

The Most Important Elements

Just as the human body doesn't have a single "most important" organ, several elements are essential for the creation of life. These are among the most important elements of all living things on earth:

- hydrogen (which is denoted by its chemical symbol, H)
- carbon (C)
- nitrogen (N)
- oxygen (O)

Whether in the air we breathe, the food we eat, or the materials that make up the physical structures of the human body, without these elements humanity would not exist. What makes these elements so essential to the formation of life is their ability to interact with other elements and then organize them into important molecules (matter that is composed of more than one atom) or compounds (molecules composed of two or more different elements). They can do this because of their subatomic structure and particles.

Anatomy of a Word

molecule

A molecule is a piece of matter consisting of more than one atom. A molecule can be made up of atoms that are all the same element (such as a molecule of oxygen) or it can be made up of different atoms, meaning that a molecule can be a compound (such as a molecule of water, which is a combination of hydrogen atoms and oxygen atoms).

Subatomic Particles

All atoms are made up of three basic subatomic (i.e., anything smaller than an atom) particles:

- protons (which have a positive electrical charge)
- neutrons (which have no charge)
- electrons (which carry a negative electrical charge)

The number and organization of these particles dictates whether an atom will readily interact with any other atom—and also defines what type of atom it is. If an atom has only 1 proton, it must be a hydrogen atom.

Positively charged protons are found in the nucleus of the atom.

Where do atomic numbers come from?

The number of protons present in an atom is the atomic number for that element. For example, carbon has an atomic number of 6 and oxygen has an atomic number of 8, which means carbon has 6 protons and oxygen has 8 protons in the nucleus.

Another particle found in the nucleus of an atom is the neutron. While neutrons don't contribute any charge to the atom, they do contribute to the mass of the atom. Therefore, the atomic mass of an atom is the number of protons *and* neutrons present in the atom. So while carbon has an atomic number of 6 (6 protons), it has an atomic mass of 12 (which means there are also 6 neutrons in the nucleus).

However, while the nucleus is populated, there is an unequal charge for the atom. As with most things in the universe, atoms seek balance. To obtain this balance, atoms have negatively charged particles that orbit around the nucleus. These are called electrons. It is the electrostatic attraction between the electrons and the protons that keeps the electrons spinning in orbit around the nucleus, much like the moon is held close to the earth by gravity. In fact, to find a natural balance, atoms will have the same number of protons as electrons, leaving the atom with an overall net neutral charge.

Electrons, however, are not restricted to a single location, such as the nucleus. They are found in orbitals (shells) around the nucleus. An atom can have many orbitals. In illustrations, these will often be drawn as concentric circles with the first being closest to the nucleus. The first orbital of any atom (that is, the orbital closest to the nucleus) can contain up to 2 electrons. After this orbital is filled, if the atom has more electrons, they will be packed into the next orbital, which can contain up to 8 electrons. Once the next orbital is filled (if there are more electrons), then they are packed into the next and so on. All orbitals after the first can contain up to 8 electrons.

Orbitals by the numbers

For carbon, with an atomic number of 6 (meaning 6 protons and thus 6 electrons), 2 of the electrons will be in the first orbital and the remaining 4 will be in the second (and outermost) orbital.

With this basic understanding of atoms and subatomic particles, you can better understand how atoms will combine together to form molecules and compounds.

The *real* building blocks of matter

The fact that subatomic particles exist is why scientists cry out in despair when people say atoms are the building blocks of matter. Particles such as protons are smaller than atoms, and for many years scientists thought *they* were the building blocks of matter . . . until someone discovered quarks, which have itty bitty charges and combine to form protons and neutrons. No one has actually seen a quark, but experiments show they must exist. Thus it is quarks that are actually the building blocks of matter (and will continue to be until someone finds something smaller).

Periodic Table of Elements

In order to show the relationship of the various elements, scientists have arranged them into the periodic table of elements, which you probably remember from high-school chemistry class. The table of the elements begins with the element that has an atomic weight of 1 (hydrogen) and goes to—well, that depends on the table you're looking at. There are 114 confirmed elements and several others are suspected to exist, such as 118 (ununoctium, a synthetic element no one knows much about, sort of like that weird neighbor down the street). Ninety-eight elements occur in nature; the others are only found in labs (where they are synthesized).

Each entry in the table includes the element's atomic number and its chemical symbol. Some tables may also show the atomic mass. Color coding is often used to indicate groups of elements that share similar qualities.

CHEMICAL BONDS

How Atoms Stick Together

Atoms sometimes create connections (bonds) with other atoms, allowing them to form relationships we call molecules or compounds. Sometimes these bonds are long-lasting, and other times they are shorter-lived than that thing you had with that drummer back in high school. Bonds between atoms are generally created by the attraction of opposite charges. For that reason, if an atom has an outer shell (orbital) that is already filled with electrons, it is unlikely to form molecules or compounds with other atoms/elements.

However, if the atom has room in its life (outermost orbital), it is more receptive to a bond. A bond can be accomplished by the atom either giving or receiving electrons from other atoms, or by sharing electrons with other atoms.

Ionic Bond

An ionic bond is when 2 atoms form molecules by giving up or taking electrons from others to complete their outermost orbitals. The classic example is the compound salt (sodium chloride, NaCl). Na (sodium) has a single electron in the outermost (third) orbital. That is one lonely electron. To fill the outermost orbital, sodium could recruit 7 more electrons from other atoms, but that would be a lot of work and impractical and is totally against the law in several states. Therefore, Na gives up the single electron and leaves the complete second orbital filled with 8 electrons, a very stable arrangement. However, now this atom has 10 electrons and 11 protons. This imbalance between protons and electrons yields an ion. In this case, the sodium ion with 10 electrons has an overall positive charge.

On the other hand, chlorine (Cl) has the dilemma of needing a single electron to complete its outermost shell. With an atomic number of 17, there are 7 electrons in the third orbital of chlorine and there is room for 8, making it a natural partner for sodium (and it didn't even have to join an online dating service). Sodium gives up its electron to chlorine, which then uses the electron to complete its shell. Since it now has 1 more electron than proton, it has become a chloride ion with an overall negative charge.

This is where the bond happens. The positive charge of the Na^+ ion is attracted to the negative charge of the Cl^- ion, and the two will form a moderately strong chemical bond to create NaCl, or salt.

Anatomy of a Word

ion
An ion is a charged atom that has an unequal number of electrons and protons. An ion can be positively or negatively charged depending on whether it has fewer electrons than protons (positively charged) or more electrons than protons (negatively charged).

Hydrogen Bond
Hydrogen bonds are formed when atoms share electrons unequally in compounds. Water is the classical example of this type of bonding. Hydrogen has an atomic number of 1, so its shell is half full. Oxygen, with an atomic number of 8, lacks 2 electrons from filling its outermost shell. Thus, oxygen will share an electron with 2 hydrogen atoms, which will complete the outer shells of all three members of this compound, creating H_2O, or water. (The subscript 2 on the chemical abbreviation for hydrogen indicates that there are 2 atoms of hydrogen in the compound.)

However, with more protons in the nucleus of oxygen, the shared electrons will spend more time around that nucleus than around either hydrogen nucleus. This imbalance will create a slight negative charge on the oxygen side and slight positive charge on the hydrogen arms. This polarization of charge will cause water molecules to be attracted to each other. In this way, water will adhere to itself. This type of bond is the weakest of the three chemical bonds. It is also the type of bond that holds two strands of DNA (genetic code) together in chromosomes, which are the instruction books that tell an organism how to be what it is supposed to be.

Covalent Bond

The strongest of the chemical bonds, the covalent bond is when a molecule or compound shares electrons equally. Carbon, the foundation atom of organic molecules, is well adept at this type of bond formation since its atomic number of 6 means it needs 4 electrons to fill its outermost shell. Because of this, carbon can form 4 single covalent bonds with other atoms.

What is an example of a compound with a covalent bond?

A great example of a compound with a covalent bond is the basic structure of an amino acid. Amino acids are organic compounds that combine to form proteins (which are necessary to create tissue, organs, hair, skin—you name it, it needs an amino acid to help build it). The carbon is the central atom in an amino acid, onto which four components attach, each using one of the available bonds: a carbon group, a nitrogen group (called an amino group), a single hydrogen atom, and a fourth group, the structure of which changes from amino acid to amino acid. This variable side group (sometimes referred to as a side chain) is called an R group.

pH: Ions, Acids, and Bases

A pH measurement tells you whether a substance is an acid or base. An *acid* has a low pH and will release hydrogen ions (under certain circumstances) and a *base* has a high pH and will release hydroxide ions (under certain circumstances). Vinegar is an example of an acid. Baking soda is an example of a base. Acids and bases react if combined. If you mixed vinegar and baking soda together, you would produce a gas (which creates those bubbles and that hissing noise).

A mixture's pH is essentially a measure of its hydrogen ions. If a substance has molecules or compounds that will yield a large number of H^+, the substance, based on a mathematical logarithm, will result in a lower pH number and be considered an acid or acidic solution (pH < 7.0). Conversely, with lower H^+ concentrations, the pH will be above 7 and considered a base (also called a basic solution or alkaline). This standard is set internationally using known materials, such as pure water (pH of 7.0).

Most living organisms can survive only within a small pH range. Change the pH range and you'll create a reaction, and change (or kill!) the organism. Cells can change pH through the process of metabolism (think of your muscle cells creating lactic acid when you exercise hard).

Your slightly alkaline bodily fluids

The pH of plasma and body fluids is approximately 7.3–7.4. This is on the alkaline side of neutral (7.0). It is called the physiological neutral point. Your body will try to maintain this level of pH through various cellular processes.

ORGANIC COMPOUNDS: CARBOHYDRATES AND PROTEINS

It Does a Carbon-Based Life Form Good

Most life on the planet is carbon-based. The chemistry associated with carbon-based living organisms is called organic chemistry and focuses on carbohydrates, proteins, lipids, and nucleic acids. These are called "organic" compounds and are used to build everything you need to run that marathon you're entered in this weekend—from the lungs you use to breathe to the energy you use to power your stride.

Carbohydrates

Also known as "sugars," carbohydrates (or saccharides) play a major role in energy conservation, transport, transfer, and storage. Plants capture energy from the sunlight and use it to assemble carbon molecules into carbohydrates. When you eat a plant, your body breaks down these complex molecules into individual CO_2 molecules and recovers the energy generated from the breaking of the bonds to be used elsewhere in the body. In your body, energy can be stored either as fat or as long chains of carbohydrates (polysaccharides).

A single carbohydrate molecule is often referred to as a monosaccharide with a typical chemical composition of $(CH_2O)_n$ where n is at least 3. Thus, $C_3H_6O_3$ is the simplest of monosaccharides (it is called glyceraldehyde). Glucose, one of the most important energy-bearing monosaccharides, is $C_6H_{12}O_6$.

Disaccharides are composed of 2 monosaccharides. Sucrose is a disaccharide composed of glucose and fructose, another important monosaccharide for metabolism. The common name for sucrose is table sugar.

Beyond disaccharides

Oligosaccharides are composed of between 3 and 9 monosaccharide units. Polysaccharides (a saccharide of more than 1 monosaccharide) may be much longer.

Saccharides provide energy storage. Glucose is polymerized (the process of linking together molecules) into glycogen, which is stored intracellularly (inside the cell) in muscle and liver cells to be broken down in times of high energy needs and low blood glucose levels, such as when you oversleep and end up running out the door before you even have a chance to eat breakfast.

Proteins

Protein molecules serve as structural elements both inside and outside of the cell, as anchoring molecules to hold cells in place, as adhesive molecules to allow cells to move throughout the body, and as enzymes that facilitate much of the metabolic activity of the cell.

Amino Acids

Amino acids link together to form proteins, using a special linkage called a peptide bond. Because of this, proteins are often called polypeptides. There are twenty types of amino acids, each with a different structure. The final shape and function of a protein is determined by its amino acids. Since the only variable part of an amino acid is the R group, this is the portion of the amino acid that will confer different physical and functional properties to the protein. For instance, several amino acids, such as valine and isoleucine, have hydrocarbon (molecules consisting of only carbon and hydrogen) R groups. These groups will be neutrally charged and will not interact with charged (also called polar) molecules, such as water. Thus, these amino acids are said to be hydrophobic ("afraid of water") and are often present in regions where a protein will span the plasma membrane (which is also a hydrophobic region of fatty acid hydrocarbon chains). Other amino acids are hydrophilic (attracting water). Some are acid; others are base.

Amino acids and protein folding

The type of amino acid in the protein will have an impact on the shape and folding of proteins into their final structure. A string of amino acids only becomes functional when it folds into a protein. For instance, glycine has the smallest of the R groups, with only hydrogen present. This will allow the protein to fold easily since there isn't a large R group to physically get in the way.

Protein Structure

Methionine will always be the first amino acid in a protein since it is also the sequence (in the RNA) that signals the start of protein formation. While the protein sequence is usually written in a straight line and is considered the primary structure of the protein, proteins are flexible and typically fold back upon themselves into one of two patterns:

1. in side-by-side runs of the protein, forming beta pleated sheets (sheetlike regions of the protein)
2. twisted around neighboring regions of the protein, forming spiraling tubes called alpha helices

Each of these folded patterns, which form the secondary structure of the protein, is held in place by hydrogen bonds between amino acids.

As the protein folds, other amino acids may become closer to each other and form bonds. Cysteine, for example, is an amino acid with a sulfate group. When it is next to another cysteine it may form a disulfide bond. In this way, large loops of protein are held in place. These formations within the protein are called the tertiary structure of the protein.

Lastly, separate protein units may be held together by bonds into large protein aggregates. This quaternary structure (meaning the combination of 2 or more chains that form a final structure) is illustrated in the hemoglobin molecule. Adult hemoglobin (the protein responsible for oxygen transport) is formed from 4 subunit proteins, bonded together in the large single molecule that moves oxygen throughout the blood stream.

ORGANIC COMPOUNDS: LIPIDS AND NUCLEIC ACIDS

Fatty Acids and the Code Talkers

Carbohydrates and proteins may get all the press but two other organic compounds, lipids and nucleic acids, are fundamental to cellular biology—and therefore to human life.

Lipids

Lipids are hydrocarbon molecules used in plasma membranes and for energy storage. Because they are composed of neutrally charged hydrocarbon chains, they are hydrophobic.

What keeps oil and water from mixing?

Oil is a hydrocarbon (hydrophobic substance) and will not interact with charged (polar) molecules, such as water.

Saturated versus Unsaturated Fatty Acids

Fatty acid chains are polymers of hydrocarbons, which are attached to a carboxylic acid (a compound in which a carbon atom is bonded to an oxygen atom, such as COOH-, making it a weak acid). The carbons can be attached to each other via a single bond, with the remaining bonds completed with hydrogen molecules. This would generate a saturated fatty acid in which all of the bonds of carbon are occupied by additional atoms. A straight linear fatty acid chain is generated when the carbon bonds are saturated.

However, a double bond may be present between carbons (and therefore one fewer hydrogen on each of the adjacent double-bonded carbons). This is an unsaturated fatty acid and will have a bend at each of the double-bonded regions. Any more than one double bond in a single fatty acid chain will result in a polyunsaturated fatty acid.

Saturated and unsaturated fatty acids

Eating foods that are high in saturated fats can increase your cholesterol (a waxy substance that can clog your arteries), possibly affecting your heart health. Most saturated fats come from animal sources (including meats and dairy). Unsaturated fats are considered healthier for your heart.

Phospholipids

Phospholipid is the main constituent in membranes within a cell, including the plasma, nuclear, mitochondrial (the mitochondria is the energy-producing structure in a cell), and vesicular (vesicles are structures used for storage and transport) membranes. Basically, 2 side-by-side fatty acid chains are attached at one end to a glycerol molecule. One end of the molecule is charged, making it hydrophilic, and the other end, composed of fatty acid chains, is not, making it hydrophobic. This duality is called amphipathic.

Triglycerides

Triglycerides are the storage form of energy in the body and typically are referred to as "fat." This material is stored in fat cells called adipocytes and can be recruited when your body requires the use of a lot of energy. Releasing double the amount of energy as compared to glucose, triglycerides take longer to release energy than carbohydrates because it takes the cells longer to break down the fat and place it into the blood stream.

As the name implies, these molecules are composed of 3 (tri-) fatty acid chains attached to a glycerol molecule.

Sterols

In humans, the principal sterol (a type of fat) is cholesterol. Despite all the bad press, your body cannot function without cholesterol. It plays a critical role in the proper spacing of the plasma membrane, which gives stability to the membrane. Additionally, hormones such as estrogen and testosterone are derived from cholesterol, and are crucial to proper body function.

Nucleic Acids

Nucleic acids are necessary for each and every cell of the body. Nucleic acids exist in two forms, deoxyribonucleic acid (DNA) and ribonucleic acid (RNA). These linear molecules are the repository of genetic information (DNA) and copies of that information with which proteins are built (RNA).

DNA

Described as a double helical molecule, DNA has 2 chains held together by hydrogen bonds that can easily be separated for either DNA replication (during cell division) or RNA synthesis and transcription (the conversion of genetic information into proteins that carry out the directions).

DNA strands are composed of a few basic units. First, a sugar molecule will form part of the backbone of the strand of nucleic acid (for DNA that sugar is deoxyribose, hence the "D" in DNA). The other portion of the backbone is a phosphate group, which will link the sugars together into a long strand.

Also attached to the sugar is a nucleobase of either a purine or a pyrimidine. The purines of DNA are either adenine (A) or guanine (G) and the pyrimidines are thymine (T) and cytosine (C). The hydrogen bonds between the nucleobases are what hold the 2 DNA strands together into the double helix. The bases are always paired together: A-T and G-C. Because of their structure, the A-T pair is held together with 2 hydrogen bonds and the G-C pair with 3 hydrogen bonds. Thus, the G-C pair requires more energy to break than the A-T set. This becomes important for DNA replication and RNA synthesis (discussed in later sections).

Purines and pyrimidines—what's the difference?

Both are nitrogenous bases. A purine has 2 carbon-nitrogen rings while a pyrimidine has a 1 carbon-nitrogen structure. They have similar functions.

RNA

Ribonucleic acid (RNA) is similar in its structure to DNA with some important differences. As the name implies, the first difference is the sugar used. For RNA, that sugar is ribose. Additionally, RNA will be synthesized as a single strand rather than a double strand. Lastly, while G, C, and A are found in RNA, T will not be present. Rather, uracil (U) is used instead.

IMPORTANT INORGANIC COMPOUNDS

The Night of the Living Dead

While much of the focus in the study of anatomy is on the structure, function, and metabolism of organic molecules (molecules containing carbon), some inorganic compounds are essential for human existence and for life in general.

One way to think of the difference between organic and inorganic molecules (other than organic molecules being carbon-based) is that organic molecules are generally synthesized by living organisms, whereas inorganic molecules are not (they are usually produced by other means, such as geologic processes).

Water

While most life on the planet is carbon-based, water, which is not carbon-based, is a compound without which life would not be possible. The human body, for instance, is considered to be composed of between 50 and 65 percent water.

This water exists within your cells (about two-thirds of your water content) with the remainder outside of the cells in your tissues or blood stream. Your brain is about 85 percent water while your bones are about 10 percent water.

Water is the universal solvent because it is composed of polar molecules—that is, molecules that contain a charge—that are capable of ionizing many molecules (e.g., NaCl). A solvent is used to form a solution when another substance is dissolved in it. A universal solvent is one that can dissolve a wide range of substances. Lemonade is an example of a solution. So is saline (which is water in which NaCl has been dissolved). Both of these solutions use water as the solvent. Water in the human body is used as a solvent for elements such as chloride. Proteins and other molecules also use water as a solvent.

Anatomy of a Word

ionizing

Ionizing means any process that changes a neutral (noncharged) atom (or molecule) into one that carries a charge.

Water is an essential substrate for many processes in the human body. A substrate is a molecule on which enzymes act to catalyze, or cause, biochemical and other essential reactions. So water is the basis upon which many biological processes take place.

The connection between water pressure and blood pressure

In the blood stream, the amount of water pressure has a large impact on blood pressure and heart activity. The kidneys respond to changes in the amount of water and water pressure in the body. For example, the kidneys will excrete more water and salt if the blood pressure increases to help reduce it.

The body uses water for many functions, including:

- regulating body temperature
- lubricating joints and moistening tissues
- flushing waste (preventing constipation and reducing demand on the kidneys)

- aiding in the transport of materials and gases in the blood stream
- dissolving molecules (such as minerals) so the body can use them

The Role of Salts

Your body uses a number of inorganic compounds in the form of salts. (Calcium phosphate salts, described in the following section, are an example.) A salt is an ionic compound formed when a base reacts with an acid, which means it has net neutral charge despite being made up of charged parts (positively charged ions and negatively charged ions).

Because of the properties of salts, in the body they can dissolve to become electrolytes (free ions that are conductive), making it possible for them to carry an electrical charge through a solution. Sodium chloride, for example, helps transmit neural impulses and is necessary for your muscles to contract. It is also used to aid digestion and to help regulate the amount of fluid in your body.

Calcium Phosphates

Calcium phosphates, a type of salt, make up much of the inorganic material in the bones and teeth. These body parts are essential for support, movement, and eating, but they also play the important bodily function of storing calcium as phosphates.

Calcium is an essential ion for muscle contraction, nerve signaling, and protein activation, among other activities. If blood calcium levels decrease, calcium can be recruited from storage in the bones to maintain homeostasis of cellular activity. Endocrine glands secrete hormones that closely regulate blood calcium levels.

Acids and Bases

Your body uses acids and bases for a number of functions as well. Your body produces hydrochloric acid to digest food in the stomach. However, the hydrochloric acid must be neutralized once it is mixed with food and leaves your stomach or it would destroy other tissues, so your body produces a base, bicarbonate, to reduce the acidity.

Your body also produces buffers that can make small changes in the base or acidity of a substance in your body to help keep your bodily fluids the proper pH.

Minerals

Your body needs other inorganic substances in the form of minerals to function. Minerals are naturally occurring solids that help your body with various processes. For example, iron helps bind oxygen to red blood cells and transport it throughout your body. Someone without enough iron in her body will be anemic, and suffer from fatigue and occasionally serious, life-threatening disorders. Other minerals are used to create hormones and to regulate your heartbeat. Minerals your body needs for proper functioning include:

- magnesium
- manganese
- iodine
- zinc
- potassium
- fluoride

COMPONENTS OF A CELL

The Secret Life of Cells

Most cells in the human body consist of several organelles, units within the cell that have a specific function. Organelles include the membrane, cytoplasm, nucleus, endomembrane, and mitochondria.

Membranes

The cell membrane or plasma membrane forms the boundary between the inside and outside of a cell. It consists of both protein and lipid molecules in varying ratios depending on the cell type. Typically, there are 50 lipid molecules for each protein in the membrane. However, since proteins are much bigger than lipids, proteins make up 50 percent of the mass of the membrane. These molecules are arranged in two opposing sheets, creating a bilayer of lipid and protein. One layer faces the outside of the cell, or the extracellular surface, and the other faces the interior, the cytosolic surface.

The main lipid type in the membrane is a phospholipid, a molecule made of a charged phosphate group attached to a base molecule (either a glycerol or sphingosine). Because of the negative charge of the phosphate, one end of the phospholipid layer can interact with water molecules, which are also charged, or polar. At the other end are fatty acid hydrocarbon chains that create a nonpolar (hydrophobic) area. Just as oil and water don't mix, this part repels water or charged molecules and makes an effective filtration barrier, called a semipermeable membrane.

One role of cholesterol in cell structure

Cholesterol is abundant in the plasma membrane and serves to regulate membrane fluidity. Like phospholipids, a cholesterol molecule has a hydrophilic head and a hydrophobic tail, meaning the cholesterol molecule can align with the phospholipid and create a more rigid structure.

Proteins embedded in the phospholipid bilayer may be associated with either or both of the surfaces of the membrane. This membrane-spanning arrangement enables the proteins to serve many cellular functions. They transport material into or out of the cell and provide membrane attachment for stationary cells or adhesion for migratory cells. Many proteins in the membrane are receptors that recognize chemical signals and relay those signals to the inside of the cell to alter cellular activity.

Free-floating proteins

Proteins in the membrane are not locked in place. They may float throughout the membrane, spin, or flip horizontally.

Cytoplasm
Effectively separated from the extracellular environment by the plasma membrane, the inside of the cell is the location of most metabolic activity. The cytoplasm houses the workshops of the cell. Here, incorporated material is broken down, new proteins are generated, and new phospholipids are produced.

Nucleus

The nucleus, which contains the DNA (genetic code for the cell), is positioned in the center of the cell. The nucleus turns copies of the DNA code into RNA, which is used in the cytoplasm to make proteins.

The nuclear membrane consists of the same materials as the plasma membrane. But the nuclear membrane is composed of 4 phospholipid layers (2 bilayers) with a perinuclear space between.

The most prominent structure within the nucleus is called the nucleolus and is made up of proteins and nucleic acids. Here, rRNA (ribosomal RNA), required for protein production, is synthesized and prepared for transport outside the nucleus.

Anatomy of a Word

ribosome

Ribosomes are structures found in all living cells, responsible for producing most of the proteins that are created in an organism.

Protein complexes within the nuclear membrane regulate transport of materials into and out of the nucleus. While small water-soluble molecules may pass unimpeded, larger molecules must be assisted to get from place to place. The "helper" is a cotransport molecule, which must bind to the "cargo" molecule to allow the transport. The helper that moves molecules into the nucleus is called importin. Exportin moves molecules out of the nucleus.

Endomembrane System

Many of the membrane-bound organelles of a cell are either physically or functionally connected or both, and are thus grouped together as part of the endomembrane system. These include:

- nuclear envelope (the membrane surrounding the nucleus)
- endoplasmic reticulum
- Golgi apparatus
- vesicles
- plasma membrane

Endoplasmic Reticulum

The endoplasmic reticulum (ER), which plays a vital role in protein and lipid production, is made up of large folded sheets of membranes that occupy vast expanses of the cytoplasmic compartment. ER comes in two types:

- Rough ER (rER) is covered with ribosomes, the organelles for protein synthesis, which give the ER a rough appearance.
- Smooth ER (sER) does not contain ribosomes and is the site of lipid synthesis.

Material of the ER is transported in membrane-bound spheres called vesicles that move toward and fuse with the membranes of the Golgi apparatus.

Golgi Apparatus

The Golgi apparatus, a cell structure made up of flat sacklike layers, is essential for sorting proteins and packaging them for specific targets. As the vesicles fuse together on the incoming side, they form a new layer termed the cis face of the Golgi. Like an assembly line, these layers are moved higher and higher in the stacks of the Golgi as new cis layers are added. At the opposite side of the stacks, the trans face, or last layer, breaks up into transport vesicles and shuttles material to its target.

Some materials will be shipped to the plasma membrane, while others will be placed into vesicles going to other membrane-bound organelles, such as the mitochondria.

Vesicles

In addition to the transport vesicles of the endomembrane system, other vesicles are essential for proper cellular function. Lysosomes are spheres of enzymes that break down proteins, carbohydrates, or fats. Peroxisomes contain hydrogen peroxide and are prominent in liver and kidney cells where the hydrogen peroxide detoxifies ethanol and breaks down fatty acids.

Mitochondria

The mitochondria is a structure that creates the energy used by the cell. It consists of a double membrane system, much like the nuclear membrane. Like the nucleus, mitochondria also possess DNA. The mitochondrial genome encodes for over 30 genes whose products play essential roles in metabolism and energy production.

Shaped like a capsule, the outer mitochondria membrane is flat over the surface of the organelle while the inner membrane is folded (to increase surface area) into sheets, which are called cristae. Proteins in the inner membrane create an electron transport system where protons or hydrogen ions (H^+) are transported from the interior of the mitochondria, called the matrix, to the intermembrane space between the inner and outer membranes. The energy of the flowing H^+ is used to produce ATP (adenosine triphosphate), the molecule that all cells of the human body use for energy.

MOLECULAR TRAFFICKING

Cellular Customs and Border Patrol

Transport of material into and out of a cell can be passive, where no energy is used, or active, where the process must be helped through the expenditure of energy. In passive transport, molecules move from areas of high concentration to areas of low concentration.

Diffusion and Osmosis

Simple diffusion is a process by which molecules in areas of high concentration spread out into areas of low concentration. Spritz some air freshener in the trash can and pretty soon the whole kitchen will have a lovely lilac scent. This is an example of diffusion.

You can think of the process of diffusion as a lot like riding a bicycle downhill. The only energy required was what you used to get up to the top of the hill. Afterward, it's simply letting gravity coast you downhill. For cellular transport, the uphill push is the creation of a high concentration of molecules. This buildup of molecules might happen when, for example, food is digested into the nutrients your body needs to function. The nutrient molecules pile up. This process requires energy. But the distribution of the piled-up molecules—having them topple downhill (so to speak)—doesn't (necessarily) require additional energy.

Thus, the top of the hill is the area of high concentration and the bottom of the hill is low concentration. (If you have a stack of molecules at the bottom of the hill, the molecules at the top of the hill wouldn't have any place to go, so diffusion could not occur.)

Oxygen and carbon dioxide freely diffuse through the membrane during respiration.

Anatomy of a Word

respiration

Respiration is the process of bringing oxygen into cells and getting rid of carbon dioxide (a waste product) from cells.

As the cell does its work, it uses up its oxygen, converting it to carbon dioxide. The carbon dioxide piles up until the concentration is high enough that it can topple downhill (that is, through the cell membrane and out into the wild where it can be carried away). As the carbon dioxide builds up inside the cell, oxygen builds up outside the cell. Once the carbon dioxide moves out of the way, the high concentration of oxygen outside the cell can tumble inside the cell, evening out the distribution of oxygen molecules. And the cycle continues to infinity and beyond, or at least for a very long time.

Water is also capable of freely diffusing through the plasma membrane; however, the diffusion of water is termed osmosis.

Water molecules tend to dilute materials to an equal extent. If water on one side of a membrane has more solute added, such as sugar or salt, water will flow from the side of less solute (lesser concentration, or hypotonic) and into the side rich in solutes (hypertonic) in order to attempt to equalize the concentration of water and stuff on both sides of the membrane.

Carrier-Mediated Transport

Molecules that can't diffuse through the plasma membrane (because they are too large or because they carry a charge) are transported through the membrane via protein channels. The molecule is still moving from an area of high concentration to an area of low concentration.

Glucose and charged ions such as sodium are among the molecules and ions that must use a protein channel for diffusion into or out of a cell. No energy is used since this is still diffusion. The only difference is the specialized tunnel through which these molecules can diffuse.

Active transport is a type of transport that is distinct from passive diffusion in that molecules are actively moved in and out of the cell, often with the help of transport proteins. The molecule in need of moving adheres to a transport protein, which brings it where it needs to go and releases it. This type of facilitated diffusion is still a matter of moving molecules from an area of high concentration to an area of low concentration.

However, sometimes molecules need to be transported from an area of low concentration to an area of high concentration (against their concentration gradient). This is the opposite of diffusion, and it requires transmembrane proteins to carry the molecules plus a lot more energy to make the trip. (This is more like riding your bike up hill than it is coasting downhill.)

Active transport protein channels bind the transport molecules and use energy from ATP to change their shape in such a way as to move the molecules across the membrane and against the gradient. The final step of the active transport process is to reset the channels so that the next active transport cycle may begin.

Sometimes a cell uses membrane vesicles to transport molecules into and out of a cell. In this case, instead of using a transmembrane protein as a sled, the molecule is encased by a pocket in the cell membrane (creating a vesicle) and moved into or out of the cell.

- Endocytosis is the process of using vesicles to transport molecules into a cell.
- Exocytosis is the process of using vesicles to transport molecules out of a cell.

CELL GROWTH AND REPLICATION

Time for "The Talk"

Cellular growth and division is a simple fact of life itself. Controlled at the molecular level and via secreted materials, growth and division are usually maintained with precision, unless a cell or group of cells begins to divide out of control, which leads to the formation of a tumor.

Cell Cycle

A typical cell will spend the majority of its cell cycle—the process of growth and replication—providing an essential function to the tissues and organs where it resides. This stage of the cell cycle is called interphase. In this beginning stage, the cell grows to its final size and may remain in this static, functional state until it prepares to divide again. At this stage, called mitosis, the cell divides its chromosomes and nucleus. The cell cycle finishes with cytokinesis, the division of the cytoplasm, resulting in 2 daughter cells identical to the single parent cell from which they were produced. The resulting cells begin the process of interphase and go through the cell cycle all over again.

Interphase: G1

Once mitosis and cytokinesis are complete, each daughter cell enters the first phase of interphase: the gap 1 (G1) phase. Here, most cells increase in size, replicate essential organelles, and move the nucleus more toward the center of the cell. At the end of the G1 phase, the cell checks to be sure that the process of replication has begun without errors. If there is a problem, cell division will be halted and the cell will attempt to repair the problem.

Interphase: S

The synthesis (S) phase follows G1 and is the period when the chromosomes are duplicated so that each daughter cell can have a complete set of chromosomes.

Interphase: G2

In the gap 2 (G2) phase, the cell begins to prepare for mitosis. During this time, the cell produces and organizes all of the structures and materials essential for mitosis. The most important point is the G2-M transition. Here, the cell size, DNA replication, and DNA damage are checked before the cell continues the process of replication.

Mitosis: M

During the M, or mitotic phase, a parent cell is cloned into 2 daughter cells. Each human parent cells possess 46 chromosomes in 23 pairs. Each resulting daughter cell will be a clone of the parent with exactly the same 46 chromosomes.

Anatomy of a Word

mitosis

Mitosis is strictly defined as the process by which the chromosomes in a cell are duplicated and separated into their own nuclei.

During prophase, the first phase of mitosis, the sister chromatids (a chromatid is one of a pair of duplicated chromosomes), which were formed during the S phase of interphase, are condensed and become more coiled in preparation for cell division. Additional changes occur in the cell during this phase, such as the nuclear membrane beginning to disintegrate.

An intermediate period, prometaphase, is often considered late prophase. During this time, microtubules called spindle fibers, which stretch from each pole of the cell toward the opposite pole, are organized. The spindle fibers serve two critical functions during cell division. First, fibers from each pole connect to each side of the sister chromatids to move the chromatids to the middle of the cell. Other spindle fibers function to push the poles of the cell farther apart in preparation for the division of the cytoplasm in the last stage of mitosis.

The most often illustrated phase of mitosis, metaphase, is the point when chromosomes become attached to the spindle fibers that were previously produced. It is recognizable because of the alignment of all 46 sister chromatids at the equator of the cell.

The shortest of the phases of mitosis, anaphase is characterized by the pulling apart of the sister chromatids into 46 individual and identical chromosomes that are being moved in opposite directions.

Anaphase continues until the chromosomes arrive at the poles, which signals the start of telophase. At this time, the chromosomes begin to rapidly relax and uncoil, the nuclear membrane begins to re-form, and many of the spindle fibers disappear. This concludes the division of the nucleus, which coincides with the initiation of cytoplasmic division, or cytokinesis.

Cytokinesis

Near the end of telophase, proteins called actin begin to form a belt that extends around the equator of the cell. As cytokinesis continues, the actin ring becomes smaller and smaller, resulting in a narrowing at the waist of the cell called the cleavage furrow. This constriction continues until the 2 resulting daughter cells are pinched away from each other into independent yet identical cells.

Meiosis

Not all human cells divide by mitosis. During sexual reproduction, sperm and egg cells must be divided in half, so that each cell contains only 23 chromosomes rather than 46. This process of division is called meiosis. Thus, while mitosis is often referred to as cloning, meiosis is termed a reduction division.

Why do sperm and egg cells contain only 23 chromosomes?

When a sperm cell containing 23 chromosomes unites with an egg cell containing 23 chromosomes, they combine to form 46 chromosomes. The resulting new individual will have the final complete amount of genetic material. If each sperm and egg cell had the full complement of 46 chromosomes, the combined genetic material would result in an individual with 92 chromosomes.

To complete this reduction of genomic material ("genome" meaning "the genetic material of a cell") the cell undergoes 2 divisions, each given a name followed by the division number (e.g., prophase I, metaphase I). In meiosis I, the 46 chromosomes are divided so that each half now has 23 chromosomes.

The second division is identical to mitosis in that the daughter cells from the first division will now have their sister chromatids pulled apart. The only difference between this division and mitosis is that the starting material for meiosis II are 23 sister chromatids (in mitosis there are 46). Thus after the completion of 2 divisions, 4 daughter cells will result.

DNA REPLICATION

Genes, Unzipped

Before a cell can divide into two daughter cells, all of the DNA of the parent cells must be copied so that each daughter gets a copy. The DNA of a cell carries all of the organism's genetic material—and instructions on how to use it.

DNA Structure

The structure of DNA is a double helix (spiral shape around a common axis) consisting of anti-parallel (running in opposite directions) strands of nucleic acid. It might be helpful to picture the DNA molecule as a ladder. Each leg of the ladder represents one strand of DNA and the rungs represent the nucleotide bases that hold the entire structure together. In DNA, the nucleotides on one strand are always paired with a partner in a very consistent manner.

DNA nucleotides

Nucleotides (the building blocks of DNA) are made up of a sugar molecule, a phosphate, and a nucleobase. DNA pairs contain only four different nucleotides, whose bases are either adenine (A), thymine (T), guanine (G), or cytosine (C).

DNA strands are made up of a repeated pattern of a sugar (deoxyribose) and a phosphate. The sugar-phosphate molecule is called a ring compound because of its shape. The sugar is a form of ribose that has lost an oxygen atom (*de-oxy* meaning "minus oxygen"). If you think of the sugar molecule as being formed in the shape of a pentagon, then there are five angles where other molecules could attach.

Imagine the base of the pentagon on the bottom and the point at the top. The first angle to the right of the top point is called 1', or 1 prime. Going in a clockwise direction, the second angle would be called 2', or 2 prime, and so on until all five angles are numbered. This helps scientists discuss where these molecules on a ring compound are attaching to each other.

In DNA, in one nucleotide, the phosphate group attaches to the sugar at 5' and links to another nucleotide at the 3'. In this way, DNA strands exist as a one-way street that runs 5' to 3' in biochemical terms. But remember that DNA strands are connected to each other (the legs on the ladder). So, while one strand (the coding strand) is read left to right as 5' to 3', the other strand of the double helix (the complementary strand) is present 5' to 3' in the opposite direction.

Preparing for Replication

Before the DNA double helix can be replicated, it must be separated into single strands. This is like unzipping a pair of pants. On the DNA molecule, helicase, an enzyme, binds and separates the nucleotide pairs, freeing one strand from its partner. Although both strands are being replicated at the same time, for simplicity the strands and their replication will be considered separately.

How does helicase work?

Helicase breaks down the hydrogen bond at the center of a segment of DNA, where the bases connect to each other. This exposes the bases so that new bases can attach. Adenine always attaches to thymine and guanine always attaches to cytosine.

Once the strands are unzipped, they no longer take a double helix shape. They are linear strands of nucleotides (denoted with the letters A, T, G, and C).

Leading Strand Replication

The so-called "complementary" strand of DNA is the "opposite" partner for the genetic code (coding strand). But it is used as a template to produce a new double helix for one daughter cell. By using each strand as a template, each daughter cell contains one new strand and one inherited strand from the parent and therefore forms the basis of semi-replicative division.

With the DNA unzipped, an enzyme called DNA polymerase reads the template strand in a 3' to 5' direction, and assembles a new coding strand in a 5' to 3' direction. For example, if the sequence on the complementary DNA strand were 3'-A-T-C-G-G-T-T-A–5', then the new coding strand would be assembled in this order: 5'-T-A-G-C-C-A-A-T-3'. Since this is also the direction in which the DNA helicase enzyme is unzipping the double helix, DNA polymerase simply follows along until the entire strand has been copied into a long new coding strand, which is called the leading strand.

Lagging Strand Replication

Although the leading strand is continuously synthesized as one long new strand, the other replication, which uses the coding strand as a template, is more complex. Instead of a strand being continuously replicated, the lagging strand is produced in discontinuous pieces.

Recall that as DNA helicase is unzipping the double helix, it moves in one direction. This works great for the leading strand replication because the new strand can be generated in the same direction that helicase is moving.

However, for the other template, the new strand is generated in the opposite direction. This causes the new lagging strand to be generated in pieces. As soon as an area of single-stranded DNA is exposed, a new strand is being generated. But, in this case, the new strand is being formed in the opposite direction from the way the helicase is unzipping DNA. Thus, as helicase unzips away from the new strand segments, a gap of single-stranded DNA will be present between the new strand and the helicase. This is called an Okazaki fragment. The process of unzipping and replicating will continue producing distinct Okazaki fragments along the length of this template into a discontinuous lagging strand with gaps remaining between the fragments.

These fragments have to be linked together to form a new, continuous strand of DNA. The gluing, or ligation, of these fragments into one long continuous strand is the function of first DNA polymerase and then DNA ligase, which bind to the gap site between adjacent Okazaki fragments and ligates them all together into a complete strand.

TRANSCRIPTION AND TRANSLATION

C'est la Vie

The genetic code of a cell is housed in the nucleus as DNA. However, before this code can be read, interpreted, and used to create proteins, a copy must be made and shipped into the cytoplasm where protein synthesis occurs. The process of making this copy of the genetic code is called transcription and is very similar to DNA replication, with only a few differences.

Transcription

During transcription, nucleotides are assembled into an RNA molecule by the enzyme RNA polymerase. This enzyme, which is much larger than DNA polymerase, is capable of binding to specific sequences of DNA, unwinding the DNA, reading a single complementary strand of DNA as a template, and generating a single strand of RNA that will contain the genetic code for making protein.

Once the RNA molecule is produced, it detaches from the RNA polymerase and the DNA strands bind back to one another to reform the double-stranded molecule.

RNA nucleotides

While the nucleotides in DNA are A:T and G:C pairs, the nucleotide uracil (U) replaces thymine (T) in RNA. Thus, the nucleotide pairs in RNA are A:U and G:C.

The idea of a transcript in written language is to produce a word-for-word copy that uses the exact same language as is spoken. In the case of the genetic code and nucleotide alphabet, a strand of RNA is the transcript, consisting of the same code as found in the coding strand of DNA.

Types of RNA

Several types of RNA will be produced by transcription. Messenger RNA (mRNA) is the genetic code in RNA form. Ribosomal RNA (rRNA) will be combined with proteins to form ribosomes, which synthesize protein. Lastly, transfer RNA (tRNA) is molecules that transport amino acids in the correct order to the ribosome for assembly into protein.

Translation

When a translation of a spoken language occurs, the communication happens in a completely different language, often using a different alphabet. The same is true in cellular translation. Here, the machinery of the cell reads the genetic language of nucleotides and assembles proteins using an alphabet of amino acids. The essential components of translation are the 3 RNA molecules: mRNA, rRNA, and tRNA. Cellular translation can be considered the carrying out of the instructions in a genetic code. The code describes what to do and the translation does it.

RNA

Much like words are composed of letters, the genetic alphabet of nucleotides is arranged on mRNA into 3-letter words called codons, which are simply small units of genetic code—like syllables. With 4 different nucleotides arranged into 3-letter codons, there are 64 distinct codons that can be used by the translation machinery of the cell. While mRNA represents the code, the location where this code is read and interpreted and proteins are assembled is the ribosome.

Some RNA molecules can bind to specific amino acids on one end while recognizing codons of the mRNA on the opposite end. Just as the coding strand of DNA binds to its complementary strand via nucleotide pairs (A:T, C:G), tRNA has 3 nucleotides (the anticodon) arranged that are complementary to the mRNA codons. This means that as the ribosome slides along the mRNA molecule, tRNA molecules bind to their respective codons and, in doing so, bring amino acids into the interior of the ribosome in the order directed by the mRNA code, and therefore into the correct sequence for the protein.

The space inside the ribosome can only hold 2 tRNA molecules at a time. As the ribosome slides along the mRNA, the amino acids link together via peptide bonds, thus freeing 1 amino acid from its respective tRNA molecule and forming a growing chain of amino acids attached to the other tRNA. At this point, the empty tRNA molecule is ejected, the other tRNA and its growing chain slide into the space vacated by the ejected tRNA, and a new tRNA molecule with an amino acid enters. This assembly continues until the entire protein is finished.

Codons and anticodons

Given an mRNA codon with the nucleotides A-U-G, what would be the complementary anticodon of the tRNA molecule that would bind here? The anticodon that would recognize A-U-G would be U-A-C. Remember that there are no T nucleotides in RNA.

Codons

Although there are 64 possible codons, only 20 amino acids can occur in nature and be assembled into proteins. Does this mean many of the codons are irrelevant? The answer is no. Multiple codons can be recognized by the same tRNA. In addition, multiple codons encode for the same amino acid. For instance, the codons GGU, GGC, GGA, and GGG are all recognized by the tRNA for the amino acid glycine. In fact, many tRNA molecules can recognize 4 different codons.

Specific codons called start codons and stop codons signal the beginning and end of protein synthesis. The codon AUG is the one and only start codon, and signals the assembly of the first amino acid in all proteins: methionine. The codons UAA, UAG, and UGA stop translation and signal for the ribosome complex to free the newly generated protein.

ENZYMES AND CATALYSIS

Making Things Happen

Enzymes are a class of proteins made via translation. These molecules assist naturally occurring chemical reactions by making it easier to cut, modify, process, or further manipulate material into a final product. In other words, enzymes cause things to happen.

Activation Energy

Activation energy can be considered the threshold a chemical reaction must overcome in order to form a product. An enzyme lowers this threshold. Therefore less energy is required to create the product, the rate of the reaction increases, and the entire process becomes more efficient. Enzymes save energy and make the metabolic job of a cell much easier and faster.

The limits of enzymes

No enzyme can catalyze a chemical reaction that would not occur in nature. Enzymes only enable the reaction to occur faster.

Induced Fit Model

The induced fit model explains how only specific substrates can bind and be modified by specific enzymes. The model suggests that the connection between an enzyme and a substrate changes as they interact.

When a substrate molecule binds to the enzyme, the enzyme changes shape, and since the substrate molecule is bound by the enzyme, its shape is altered, too. This moves the substrate molecule into a better position to transform into the final product, thereby reducing the energy and time required for the chemical reaction to occur. You could compare this to cutting snowflakes out of a sheet of paper. If someone had to cut every side of the snowflake by hand, it would take much longer, require more energy, and likely not look very symmetrical in its final form. This would be like a chemical reaction occurring naturally. However, if the paper is first folded and then cut, fewer cuts are required, the snowflake is produced faster, and all sides will be the same. Enzymes fold and process substrates in much the same fashion.

Glycolysis

Enzymes are essential for processing organic molecules and releasing the energy stored within chemical bonds so cells can use it.

Anatomy of a Word
catabolism

Catabolism, or the catabolic process, is a process by which complex molecules are broken down into simpler ones, releasing energy as a result.

Although many organic molecules can release energy to cells, the principal molecule used to release energy is glucose. The catabolic process of breaking down glucose is termed glycolysis.

Anatomy of a Word
fermentation

Fermentation is how cells produce energy when oxygen is in short supply. In yeast, a by-product of this process is CO_2 and ethanol. In animal cells, the by-product is lactic acid.

During glycolysis, glucose is processed through 10 enzymatic steps from a 6-carbon molecule into 2 separate, 3-carbon molecules called pyruvate. These much smaller molecules can be transported into mitochondria where they support its functions. Energy is also released during glycolysis.

The reaction also causes electrons to be released. Nicotinamide adenine dinucleotide (NAD^+), a coenzyme, captures and transports electrons from one chemical reaction to another.

When stored energy in the form of ATP (adenosine triphosphate) is used by a cell, one of the phosphate groups is broken off the molecule, turning the ATP into ADP (adenosine diphosphate). During the process of glycolysis, as chemical bonds are broken and rearranged, some of the bonding energy is released and is used to convert ADP back into ATP.

Net energy gain during glycolysis

The products of glycolysis are 2 pyruvate molecules, 2 NADH (reduced NAD^+) coenzymes, and 4 molecules of ATP. In the early stages of glycolysis, 2 ATP molecules are used to prepare intermediate molecules for the next steps of glycolysis. While 4 ATP molecules are produced during glycolysis, the net production of ATP is only 2.

TCA Cycle

The next stage of energy release, also known as the citric acid or the Krebs cycle, consists of 9 enzymatic steps that break the 2-pyruvate molecules down into 6 CO_2 molecules and release the remaining energy in chemical bonds.

Each pyruvate (3-carbon molecule) must first be converted into acetyl-CoA (a 2-carbon molecule). This also results in the formation and release of a CO_2 molecule as well as the formation of another NADH coenzyme.

The further breakdown of acetyl-CoA continues as a cycle of carbon intermediates is introduced. Intermediates are highly reactive molecules with short lives and lots of energy, sort of like your first crush. The 2-carbon acetyl-CoA is combined with a 4-carbon molecule to generate the first intermediate of the TCA cycle, the 6-carbon molecule citrate.

Through 8 more enzymatic reactions, 2 more carbons will be released as CO_2, electrons will be captured in the form of 3 NADH (and 1 $FADH_2$), and an additional molecule of ATP will be generated for each acetyl-CoA molecule that enters the cycle. Thus, by the end of the TCA cycle, all carbons from glucose have been released as 6 CO_2 molecules and the energy of those bonds captured within 10 NADH, 2 $FADH_2$, and 6 ATP molecules (net ATP production remains 4 ATP). Clearly, the bulk of the energy from glucose is contained within the coenzymes and has not yet been converted into ATP for the cell. That process of ATP formation also happens in the mitochondria as the electron transport system.

Electron Transport System

The electron transport chain recovers energy contained within the coenzymes NADH and $FADH_2$ (an energy-carrying molecule that is the reduced form of flavin adenine dinucleotide). This system is similar to a hydroelectric reservoir, where the dam is the inner membrane of the mitochondria and the water is the electrons. Just as water builds up in the lake and, because of gravity, has a large potential energy to flow through the turbines of the dam to create electricity, protein complexes in the mitochondrial membrane use the energy of the electrons from the coenzymes to move hydrogen ions into the intermembrane space. This reservoir of ions then flows back into the matrix of the mitochondria through the last protein complex called ATP synthase. Just as the flow of water powers the turbine in the dam, the flow of hydrogen ions enables ATP synthase to generate ATP molecules.

Although only a net of 4 ATP molecules were produced during the stages leading up to the electron transport chain, the energy bound in the coenzymes as electrons from glucose are used to generate an additional 32 molecules of ATP. At the end of this process, 36 net molecules of ATP and 6 molecules of CO_2 are produced for each molecule of glucose.

TISSUE ORIGINS AND DEVELOPMENT

Making a Tissue of It

Tissues are collections of similar cells that define a specific layer of and relate an essential function to an organ. While the adult human body consists of over 200 different cell types, each human began his or her life as a single fertilized egg cell, which divided and gave rise to all the rest of their cells.

The Creation of Tissue

To understand how tissue fits in the organization of the human body, think of it as the level between cells and organs. Tissue is made up of cells and in turn it is used to make organs.

Stem cells and tissue

Stem cells are unspecialized cells that can, under certain conditions, take on specific cell, organ, and/or tissue functions. For example, under the right circumstances, a stem cell might be persuaded to become a neuron, helping to conduct electrical impulses throughout the body. In some tissues, stem cells work to repair damage, sort of like a corps of engineers.

Early in embryonic development, the newly (and rapidly!) dividing cells produce three distinct layers of cells from which all the cells of the body are derived. This process, called gastrulation, begins when a cluster of cells (called a blastula) organizes into layers. At first there is only an inner and an outer layer. These layers work together to produce a third middle layer. Organisms, such as humans, that have these three layers are called triploblastic.

These layers are called germ layers. Here the use of the word "germ" has nothing to do with disease-causing pathogens, but is related to the word "germination." Germ layers form during the embryonic ("germination") stage of human development, and they are the origin of all tissues and organs in the body (thus they "germinate" these structures). The layers are called:

- the ectoderm
- the mesoderm
- the endoderm

The ectoderm (*ecto* meaning "outside") refers to the tissue that makes up your skin, which is the outer covering of the body. The ectoderm also creates the neural tube that eventually becomes home to your central nervous system.

The human body is often considered a tube within a tube. In this sense, the ectoderm produces the outside tube of skin and the endoderm (*endo* meaning "internal") produces the tube on the inside. That tube is the digestive tract.

This leaves a lot of tissue and organs in between, which make up the mesoderm (*meso* meaning "middle"). Muscle, bone, blood, and connective tissue are all derived from this middle layer that is produced early in cell development.

Anatomy of a Word

organogenesis

Organogenesis is the name for the process by which the germ layers form all of the organs in the human body.

Your body has four main types of tissue:

- epithelial, which generally works to protect your body
- connective, which joins structures together
- muscle, which contracts
- nervous, which coordinates your body's movements

How does the anatomy of a tissue relate to its physiology?

Think of anatomy as "form" or "structure" and physiology as "function." They are linked together and can't really be separated. For example, muscle tissue has the property of being able to contract (anatomy), so it is used to move the body around (physiology).

Histology is the study of tissues and how they work to allow the human body to function as a whole. Scientists often refer to the *parenchyma* of an organ, which is the tissue of the organ that performs its essential function, and the *stroma*, which is other tissue in the organ, such as a vein or a nerve, which does not perform the essential function. A true understanding of how an organ works (or why it may not be functioning correctly) requires attention to both the parenchyma and the stroma.

EPITHELIAL TISSUE

I Feel Your Pain

Epithelial tissue, one of the four basic types of human tissue, covers a surface or forms the lining of a hollow organ. The surface of your skin is made of epithelial tissue, as is the inside lining of your stomach and intestines. In fact, the inside of your body cavity is covered with a thin layer of epithelial cells. The function of these covering cells is to protect and to provide a watertight barrier to keep material out (skin preventing pathogens from entering) or keeping material in (lining of the stomach keeping hydrochloric acid from damaging other areas of the body).

Types of Epithelial Cells

Epithelial cells are classified, in part, based on their shape. Cells can be very flat, much like a fried egg, with the nucleus of the cells bulging upward like the yolk of the egg. These flat cells are called squamous epithelium. Epithelial cells can also be cube-shaped, where the width of the side is the same as the height of the cell. These are known as cuboidal epithelium. Lastly, cells that are taller than they are wide, thus looking like columns, are classified as columnar epithelium.

Number of Layers

Epithelial tissue composed of a single layer of cells is called a simple epithelium. In simple epithelium, all of the cells in the tissue are in contact with the structure that underlies the tissue (called a basement membrane). When epithelial tissue has multiple layers of epithelial cells, it is referred to as stratified. In this stratified layer, only the cells at the bottom of the layer are in contact with the underlying tissue. In some areas of the body, epithelial tissue appears to be made up of multiple layers of cells, but closer observation shows that all of the cells are in contact with the basement membrane and it just happens that the cells are tall or have a variety of heights. This type of epithelium is classified as pseudostratified, and is easy to confuse with the truly stratified layers.

When describing the epithelium, both their shape and the number of layers are considered. A single layer of flattened cells would be called a simple squamous epithelium. Likewise, an epithelium consisting of multiple layers and having a surface layer of cube-shaped cells would be called a stratified cuboidal epithelium.

How is epithelial tissue classified when multiple layers are present?

When multiple layers are present, the shape of the cells at the surface of the epithelial tissue will be used in the classification regardless of the underlying cells.

Transitional epithelial tissue is found along the urinary tract and in the lining of the bladder. While this is a stratified epithelium, the surface cells are large and either dome-shaped (when the bladder is empty) or flattened (when the bladder is full). Often, the cells of this tissue will contain 2 nuclei, making for easy identification.

Apical Modification

The top of the epithelial cells that are adjacent to the lumen, or the hollow space of an organ, is referred to as the apical surface; these cells have membrane specializations that affect the physiological function of the tissue. One such modification on the epithelial cells lining the respiratory tract is cilia. These hairlike structures extend upward from the apical cell surface and can bend back and forth to move materials.

On cells in the intestines, the apical modifications are called microvilli. These fingerlike projections increase the surface area of the cell for greater absorption of nutrients and water.

Basement Membrane

Beneath every epithelial layer is a zone of molecules that aids in anchoring the cells to the underlying tissues in much the same way as a foundation or a basement secures a home to the ground. The basement membrane (BM) is also a transition zone where cells anchor to molecules such as laminin (a cell adhesion molecule) and other BM molecules interconnect with the underlying connective tissue, firmly anchoring the epithelial layers to it.

The BM consists of three zones, each referred to as a lamina, which is another term for layer:

* The lamina lucida is a clear layer directly beneath and in contact with the bottommost epithelial cell, containing cell adhesion molecules.
* The next layer is called lamina densa, so named because it's dark, a result of its highly compact network of type IV collagen fibers that resemble a net. This provides another anchorage point for the cells.
* The deepest layer is the lamina reticularis. In this layer, fibers from the underlying connective tissue extend upward and interconnect with the molecules of the lamina densa.

Anatomy of a Word

basal lamina

Together the lamina lucida and lamina densa compose the basal lamina. Some people confuse this with the basement membrane, but they are not interchangeable terms.

CONNECTIVE TISSUE AND MUSCLE TISSUE

Tissues of the World, Unite!

Two basic types of human tissue are connective tissue and muscle tissue. They have distinct forms and functions, but along with epithelial tissue and nervous tissue, they work together to make the human body function.

Connective Tissue

As the name implies, connective tissue joins other tissues together. It is composed of cells and molecules that function together for this adhesive process.

Connective Tissue Cells

Fibroblast is the principal cell of connective tissue. This cell deposits fibers that are found in all connective tissues: collagen, a protein resistant to stretching, thus giving the tissue tensile strength. These cells also produce elastic fibers that allow the tissues to rebound after being stretched. Macrophages are also found in connective tissues; they function as the vacuum cleaners of the body by removing pathogens and debris. Fat cells, or adipocytes, may also be present in connective tissue. These cells are mostly droplets of fats (lipids, cholesterol, or fatty acids) that are stored and released depending on the body's available energy. When fuel for the body is abundant, materials are stored as fat. In times when fuel is scarce in the blood, fat is converted into a useable form of energy.

Connective Tissue Classification

Connective tissue is classified based on the percentage of cells to fibers and how tightly packed the fibers are within the connective tissue.

- Loose connective tissue consists of widely spaced fibers and many cells migrating within the open spaces. Areas where you find loose connective tissue include the tissue around large blood vessels, beneath the epithelium of the skin, and in the digestive and respiratory tracts.
- Dense irregular connective tissue has many more fibers and fewer cells than loose connective tissue and is classified based on the orientation of the fibers. In the dermis of the skin, the collagen fibers have a swirling and disorganized pattern, so the connective tissue is called dense irregular.
- Dense regular connective tissue is made up almost entirely of collagen or elastic fibers and contains few cells. Since the fibers are tightly packed and arranged parallel to one another, this tissue is classified as dense regular connective tissue. Ligaments and tendons that connect bone to bone or muscle to bone, respectively, are resistant to stretching and are made of dense regular connective tissue.

Muscle Tissue

Muscle not only moves the body, it moves materials through the body. All muscles have one job: to contract. Muscle contraction can only happen with the sliding action of two proteins, actin and myosin. The overlapping actin and myosin molecules slide toward each other, pulling each end of the cell and shortening the muscle.

Skeletal Muscle

Skeletal muscles are muscles attached to bones. This arrangement allows the muscles, which are the engines, to move the bones, which are the levers upon which actions can occur and work can be done.

During embryonic development, individual muscle cells fuse together to create long tubes of muscle cells, which contain many nuclei. These are called skeletal muscle fibers. All of the nuclei are pressed to the periphery of the cell membrane because the middle of each muscle fiber is filled with long columns of overlapping actin and myosin molecules. The repeating and overlapping nature of the actin and myosin give skeletal muscle cells a striated appearance.

Anatomy of a Word

sarcomere

The repeating units of actin and myosin, which are arranged in series like the links in a chain, are called sarcomeres. To contract, each sarcomere shortens a small amount. When added together, all sarcomeres shortening at the same time results in the entire muscle organ shortening by as much as a few centimeters.

Skeletal muscle is the only type of muscle in the human body under voluntary control. Picking up a glass or running a race could not occur without your conscious control.

Cardiac Muscle

Like skeletal muscle cells, cardiac muscle also contains overlapping actin and myosin arranged into sarcomeres, yielding a striated appearance. However, cardiac muscle is under subconscious control and is considered involuntary. While you can increase or decrease your heart rate by your level of activity (running can increase the rate, while lying down will decrease the rate), heart rate cannot be voluntarily controlled by thought itself.

Cardiac cells are branched at many different points, unlike linear tubes of skeletal muscle. These branch points link cardiac muscle cells together in interwoven layers (laminae). Laminae allow the 3-dimensional contraction of the heart (rather than the linear contraction of skeletal muscle).

Cardiac muscle features attachment points between cells, called intercalated disks. Intercalated disks allow the muscle cells to hold on to each other tightly while contracting. These disks also contain membrane tunnels called gap junctions, which allow the cytoplasm of one cell to flow unimpeded into the adjacent cell. Thus, muscle cells joined via gap junctions contract at the same time. The heart functions as a single unit although composed of many parts.

Smooth Muscle

Smooth muscle lacks the striated pattern of skeletal and cardiac muscle. This does not mean smooth muscle lacks actin and myosin. Instead, it lacks the sarcomeric arrangement of the contractile proteins. In smooth muscle, the overlapping actin and myosin are attached to points on the plasma membrane called dense bodies, which are scattered all over the surface of the cell. This 3-dimensional pattern causes the cell to collapse upon itself when contracted.

Like cardiac muscle, smooth muscle is involuntary and may be joined together with gap junctions to form belts or bands of smooth muscle tissue. These are found in surrounding hollow organs, such as the digestive tract, urinary tract, and blood vessels, and assist in the movement of materials such as urine. For example, the pyloric sphincter is a belt of muscle between the stomach and small intestine that regulates when and if material passes from the stomach to the intestines. Contraction of this smooth muscle belt causes a narrowing and even closing of the passageway and a relaxation leads to an open passage.

NERVOUS TISSUE

That Takes a Lot of Nerve

Nervous tissue sends electrical signals from one place to another in the body. These signals can either bring information to the central nervous system from the body for processing or send information out to the body. You would trigger nervous tissue reaction if you touched a hot stove, for example. Your neurons would tell your brain that you are touching something hot (bringing information to the central nervous system). Your brain would tell your body to move your hand off the stove (sending information out to the body).

Neurons

Neurons are the signaling cells of the nervous system and come in a myriad of shapes and sizes. Generally, neurons have structures called neurites extending from the cell body, which make the cell look somewhat spiky and create the characteristic appearance of a neuron. What these neurites are named is based on the direction the signals travel. For instance, if the electrical signal travels toward the cell body, the neurite is called a dendrite. If the signal moves away from the cell body, it is an axon. Typically neurons will have many smaller-diameter dendrites and one longer, thicker axon.

Neuroglia

While neurons are the signaling cells of the nervous system, they only make up 20 percent of the nervous system. The bulk is made up of the supporting cells of the nervous system that are collectively referred to as neuroglia.

Why are neuroglia sometimes called "nerve glue"?

The word "neuroglia" comes from two Greek words that together mean "nerve glue." Neuroglial cells don't actually glue anything to anything. When they were first discovered, scientists thought they served a binding purpose, thus their name.

In much the same way as the movie stars on the screen are only a small fraction of the people involved in making a movie, the behind-the-scenes supporting members for the nervous system are the neuroglial cells (they are also called glial cells or glia). Glial cells are smaller than neurons and do not have neurites. They are not involved in actively sending or receiving signals. Rather, they support neurons by maintaining their surroundings and modulating the uptake of neurotransmitters. Glial cells also help neurons recover from injury and perform other supporting functions.

Anatomy of a Word
neurotransmitters

Neurotransmitters are the chemicals that transmit signals between neurons. Neurons have small spaces between them, called synapses, and neurotransmitters help signals pass through synapses by converting the electrical impulse of the neuron into a chemical messenger.

Myelinating Cells

One group of neuroglia insulates the axons of neurons. These cells wrap their plasma membranes around the axon 40 or 50 times, much like an electrician wrapping electrical tape around a bare wire. The term myelin refers to the area of the membrane that is tightly wrapped around the axon (to myelinate something is to wrap around it). Myelin insulates the axon, and in doing so creates a much more rapid conduction velocity of the electrical signal. For the neurons in areas of the body such as in your arms and legs, these neurons will be myelinated by Schwann cells, named after Theodor Schwann, the physiologist who discovered them. They are specific types of glial cells that exist in the peripheral nervous system and are responsible for wrapping peripheral nerves.

In the central nervous system, however, oligodendrocytes myelinate axons. Much like an octopus with many arms, these cells extend several cellular structures to axons and thus one oligodendrocyte can myelinate several different axons.

Astrocytes

Astrocytes, so-called because of their star shape, are another type of neuroglial cell. They protect the nervous system from infection by covering blood vessels with their structures and regulating the movement of materials out of the blood stream and into the nervous system. As part of the blood-brain barrier, they screen for pathogens and assist in the transportation of nutrients as well as the processing of waste.

Microglia

Microglia cells are the vacuum cleaners of the central nervous system. Known as phagocytic cells, they patrol the nervous system looking for pathogens and debris and removing them through a process called phagocytosis, or engulfing the foreign object into their cell bodies, effectively eliminating these detrimental materials before they can do any damage.

Ependymal Cells

The last type of neuroglial cell present in the central nervous system is the ependymal cell. These cells produce cerebrospinal fluid, which flows around the spinal cord and the brain and functions as a shock absorber against blows to the head and spine. Additionally, as cerebrospinal fluid is produced and as it is eliminated it creates a current that assists the blood-brain barrier; pathogens have to swim across this river of flowing cerebrospinal fluid before gaining access to the central nervous system.

Together these various cells form the tissues and organs of the central and peripheral nervous systems in the human body.

SKIN, HAIR, AND NAILS

Beauty Is Only Skin-Deep but Skin Is Pretty Deep

Skin is the largest organ in the human body. Composed of several layers, it's the external covering of the body and it serves to protect against infection and dehydration. Also present in skin is hair, which can detect touch and helps keep the body warm. To cool the body, glands in the skin produce sweat that cools the body as it evaporates from the surface.

Epidermis

While skin covers the entire body surface, all skin isn't created equal. The distinction between the skin on your arm and the skin on the soles of your feet is significant and largely based on the thickness of the epidermal (top) part of the skin.

Up to five individual layers of cells (strata) make up the top layer, or the epidermis, of the skin:

- stratum basale
- stratum spinosum
- stratum granulosum
- stratum lucidum
- stratum corneum

Let's look at each in turn. Starting at the bottom, adjacent to the basement membrane, is the stratum basale where cells divide and produce a continuous supply of new cells as the old ones are shed from the surface. The next layer as you move up is stratum spinosum, which gets its name because its cells have a spiny shape.

Anatomy of a Word

melanocyte

A melanocyte is one of the pigmented cells that give skin cells their individual color. They occur throughout stratum spinosum layer of the epidermis. They cover over and protect the dividing cells at the bottom from damage from ultraviolet radiation, kind of like an umbrella shading you on a beach.

As skin cells (keratinocytes) move higher and higher (through the process of replacing dead cells with living ones), they become cells of the stratum granulosum. These cells are filled with granules of keratohyalin, which help give skin its structure and give these cells a grainy appearance. This is the last layer of living cells.

The stratum lucidum is next. It's a thin translucent layer of the skin and is composed of dead skin cells. On top of this layer is the final layer, which is the most variable in its thickness. This stratum corneum is composed of multiple dead cell layers and is the true first-line defense barrier against infection for the body.

Anatomy of a Word

desquamation

Desquamation is the process of shedding dead skin cells. Cells are continually shed from the epidermal surface and new cells are continually made. It takes a new skin cell about 14 days to move from the bottom layer to the surface of the skin.

Thick skin is found on the palms of the hands and soles of the feet. Of the five previously mentioned layers, the stratum corneum is by far the thickest layer. It provides great protection against the friction that occurs when walking or grasping objects; however, unlike the rest of the skin that covers your body, you will not find hair follicles or sebaceous glands in thick skin. Furthermore, you will find fewer sweat glands in thick skin than in thin skin. While your feet and palms may get moist, the volume of sweat produced is much lower than what is produced in other areas of the body.

The majority of the body is covered with thin skin. In fact, the epidermis as a whole is thinner in this type of skin since it lacks two of the five layers that are found in thick skin (stratum granulosum and stratum lucidum are missing). Thin skin possesses many hair follicles and sebaceous glands that support the growing hair. Additionally, while both skin types contain sweat glands, thin skin has a much higher density of these glands to aid in cooling the body.

Dermis

The dermis is the layer beneath the epidermis that forms a transition zone between the underlying connective tissue and the epidermal layer above. It consists of dense irregular connective tissue made up largely of collagen fibers, elastic fibers, and fat tissue. Within the dermis are blood vessels that support the skin, as well as many nerve endings and receptors that detect pressure, pain, and temperature.

Hair and Nails

Human skin may also be modified into the structures known as hair and nails, which are composed of the same material that makes up the surface stratum corneum of the skin. The major difference is how compact these layers are and how they are arranged with the other dead cells in these layers.

Hair

Imagine taking the dead cells of the stratum corneum layer and rolling them into a tightly wound tube of dead cells: that is a hair. The hair follicle is simply a deep pit on the epidermis that projects downward into the deeper dermis of the skin. The other layers of the epidermis surround and support the shaft of the hair as it grows. At the deepest part of the follicle is the hair bulb. This is where the cells divide. This is also where they are nourished by blood vessels that enter into the follicle and support the growth of the hair. New layers are added to the hair and continually extrude the hair out of the follicle and onto the surface of the skin.

Nails

The surface of the skin is relatively soft, as is most of the hair on the body. However, the nails are drastically harder and consist of tightly packed layers of dead cells. What are commonly known as nails are in fact nail plates. This hardened plate of dead cells remains attached to the underlying epidermis via the nail bed, which tightly holds on to the plate to prevent the nail from falling off. At the base of the nail, you'll find the cuticle (eponychium). The cuticle is a portion of the skin's epidermis that overlaps the newly forming nail plate as it moves forward. Beneath the cuticle new nail plate material is continuously being produced and pushing the nail forward.

At the end of the nail, the plate extends beyond the tip of the finger forming a crevasse, which is great at collecting dirt, and is technically called the hyponychium. This is the part of the nail that you are most familiar with and must trim on a regular basis to prevent the nails from growing too long.

SKIN STRUCTURES AND FUNCTIONS

You Positively Glow!

Your skin uses a number of different structures to perform its functions—and it does more than just keep germs from getting in.

Sweat and Sebaceous Glands

The skin is an intact (contiguous) sheet of cells that protects the human body. Accessory structures enable the skin to cool the body and condition the skin itself. These important glands are the sweat glands and the sebaceous glands.

Sweat Glands

In ancient times, Egyptians hung wet linen cloth in doorways and windows to cool the inside of their living areas. As the water evaporated, it cut down on the heat in the air and effectively cooled the rooms. The human body uses this same approach to cool down. This is called sweating. When the surface of the skin is wetted, evaporation occurs, and as it does, heat dissipates and your body temperature is lowered.

Why some sweat smells

Eccrine sweat glands can be found over the majority of the body and produce the fluid that cools the human body known as sweat. Apocrine sweat glands can be found in greater density in the armpits and produce a different type of sweat that when metabolized by bacteria produces a distinctive and often unpleasant odor.

The secretory portion of the sweat gland resides in the deeper parts of the dermis and produces the salty protein-and-lipid-rich fluid known as sweat. This is passed through long, coiled ducts upward and through the epidermis to be released onto the surface of the skin.

Sebaceous Glands

Sebaceous glands can be found in the dermal layer of the skin and are attached to the sides of hair follicles. Sebum, a waxy secretion produced by these glands, is injected into the hair follicle, coats the shaft of the hair, and is eliminated onto the surface of the skin. This material aids in the waterproofing and lubricating of the skin and hair in mammals.

Wound Healing

An intact layer of skin is critical in preventing infectious agents from gaining access to deeper regions of the body including the blood stream. Therefore, when an injury to the skin occurs, rapid and complete repair or closure of the wounded area must be accomplished. All layers of skin are involved in the repair process.

Inflammatory Phase

The body's initial response to a wound is to minimize blood loss. Damaged blood vessels reflexively constrict to slow the flow of blood to the damaged area. At the same time, platelets, blood cells that perform a clotting function, activate and begin the process of forming a platelet plug to prevent the further loss of blood cells and to decrease the loss of blood plasma at the injury site. This plug forms the initial foundation of the full blood clot.

At the same time, the body floods the injured site with body fluid and immune cells to flush the area of pathogens and bring in white blood cells to destroy any pathogens that remain. This inflammatory process is a nonspecific means by which the body fights off any potential infection.

Proliferative Phase

As the wound site becomes filled with the blood clot and infiltrating white blood cells, connective tissue cells, or fibroblasts, migrate into the area and begin to deposit a temporary scaffolding of connective tissue molecules to fill in the open gap in the skin. This material is the start of granulation tissue; it helps close the wound and provides a foundation for the normal tissue constituents to regrow and re-form the original state of the skin.

Since the surface of your skin is epithelial and not connective tissue, those epithelial cells on the edges of the wound start to multiply and spread. They overgrow the granulation tissue. To provide nutrients to the newly growing tissue, blood vessels grow into the granulation tissue and form new circulation.

Tissue reconstruction continues as the wound is repaired and as the body attempts to close it. The blood clot starts contracting and pulls the edges of the wound closer together, resulting in a smaller area to form new tissue.

How do scars occur?

Scars occur when there is an overproduction of connective tissue proteins such as collagen, which prevents the formation of normal epithelial tissue over the wound site. If the wound is too wide or the granulation tissue grows too extensively, then re-epithelialization cannot occur. This results in a scar.

Maturation Phase

In the final weeks of wound repair, the last portions of granulation tissue will be removed, all clotting material will be eliminated from the site, and the resident constituents of the skin will be properly formed in the exact proportions and in the correct area. Following this period, most repairs of small wounds will result in skin that shows little to no sign of any injury or defect at all.

Temperature Regulation

Most people know that the skin protects your body from disease and from dehydration. However, few are aware of the role the skin plays in temperature regulation of the body.

Sweat

As discussed earlier, when you're hot, your body will sweat. As the liquid on the surface of your skin evaporates, heat is liberated from the skin and dispersed into the air, cooling the skin. In this way, the skin and the associated blood vessels that supply the skin can be thought of as a radiator or a heat exchanger between the human body and the environment.

Conserving Heat

The skin and its vascular supply also play a critical role in conserving body heat when the core body temperature begins to decrease, such as on very cold days. When exposed to the cold, the skin of the hands and face will initially flush and turn red, which indicates the dilation of the blood vessels near the skin surface in an attempt to bring more of the body's core temperature to the skin and keep the tissue warm.

This attempt to warm the body will only continue up to a point. When the body's core temperature falls enough, the blood supply to the skin is restricted and the blood is redirected to deeper (core) areas of the body. This is the body's way of sacrificing less important areas to keep the core of the body's temperature from falling too low.

SKIN DISEASES AND DISORDERS

The Heartbreak of Psoriasis

At the time of birth a baby's skin is soft, smooth, and often completely clear of defects. However, as humans age and as their skin is exposed to longer periods of ultraviolet (UV) radiation (sunlight) and environmental factors, skin is likely to exhibit localized changes in coloration or in the proliferation of cells, which can result in moles, freckles, or tumors. Other skin diseases can occur, affecting the integrity of the skin and causing potential problems with other organ systems.

Acne

Acne vulgaris is a common skin disease that often begins or intensifies during adolescence because of changes in hormones that increase oil and sebum production. It arises when hair follicles and sebaceous glands become clogged and infected. As a result, pimples or pustules may manifest and continue to occur until the underlying causes of the skin disorder are relieved.

Anatomy of a Word

comedo

A comedo is also known as a blackhead. Comedones are another common problem associated with acne. While comedones occur because of dirt or material blocking the hair follicle, they can become a whitehead or pimple if they become infected.

Blisters

Blisters are common disorders that occur on the feet because of friction from walking or running in poorly fitting shoes or on the hands from using hand tools for long periods of time without gloves. These frictional forces tear and shear the cells of the stratum spinosum. This causes a separation of cells, and the plasma from blood vessels fills these voids and raises the upper layers of the epidermis. If the friction is continued and more damage occurs to the tissue, the blister may fill with blood due to vascular damage within the layers of the skin.

Nevi

Permanent and benign colored areas of the skin, often called birthmarks, are termed nevi (nevus is singular). When this condition is caused by of a proliferation of melanocytes (pigment cells), the tissue will have a brown to black coloration. If the colored tissue is caused by a collection of blood vessels close to the skin, the tissue will have a red color and is called a vascular nevus (hemangioma), or more commonly, a strawberry mark.

Moles are a form of melanocytic nevi. Freckles (ephelides) are flattened accumulations of melanocytes in a specific area typically brought about by exposure to UV radiation.

Skin Cancer

Basal cell carcinoma, the most common form of skin cancer, is caused by out-of-control growth of the cells at the basal layer of the epidermis. Exposure to UV radiation causes damage to the DNA in these cells. Instead of dividing and creating healthy new cells, damaged and diseased cells are created. Fortunately, this is a fairly slow-growing cancer and rarely spreads to other parts of the body. That being said, it can lead to disfiguring lesions if not treated.

Melanoma, while not the most common form of skin cancer, is the most dangerous. The melanocytes lose control of their division and form colored tumors on the skin that may resemble moles (or form from moles). What makes melanoma so dangerous is the fact that it may quickly spread widely throughout the body, where it becomes difficult to treat.

Squamous cell carcinoma occurs when the most superficial layers of the skin divide out of control. These tumors may appear as red patches of flaky skin or open sores, and may bleed from or near the tumor itself. Like melanoma, these tumors are mostly caused from UV exposure and damage to the cells and can be disfiguring and dangerous if not removed before the cancer can spread.

Allergic Reactions

To prevent infection, your immune system defends itself against pathogens such as viruses and bacteria, but sometimes it defends against substances that aren't a danger to it, such as pet dander or pollen. This immune reaction can show up in your skin, causing swelling, hives (itchy red welts), and rashes (small red bumps). When the substance causing the reaction (the allergen) is removed, your body usually calms down and the reaction disappears. Severe allergic reactions can be life-threatening, although most are more moderate.

Testing for allergies

One of the ways a doctor will test for allergies is to apply the allergen or a form of the allergen to the skin (a skin patch test) or inject it just beneath the skin (intradermal test). Many common allergies are diagnosed this way.

Eczema

Eczema is the name used to refer to a group of conditions that cause red, itchy skin. Some of these conditions aren't even related to each other. These are among the most common:

- Atopic eczema is what most people think of when they think of eczema. Also called atopic dermatitis, it causes big patches of red, itchy skin. Scientists aren't sure what causes it, although they know it's more common in people with allergies such as hay fever.
- Irritant dermatitis is caused by prolonged or repeated exposure to toxins.
- Scabies, which causes red, itchy skin, is a skin infection caused by mites.
- Stasis dermatitis is a skin condition of the legs caused by poor circulation.

Psoriasis

Psoriasis is a chronic (long-lasting) skin disease characterized by thick, silvery scales and dry, itchy patches of skin. It is caused by cells building up on the surface of the skin, and is the result of a problem with the life cycle of skin cells. Periods of outbreak may be followed by periods of relative calm. Treatment focuses on stopping the skin cells from growing so rapidly and on reducing the pain and itching associated with the disease.

THE SKELETAL SYSTEM: BONE FUNCTIONS

Everyone Has a Skeleton in the Closet

The bones of the body are the structural architecture that gives the human body its distinctive shape. Without the skeletal system, you wouldn't be able to do much of anything. You would be a big pile of soft tissue, unable even to sit on the couch and be a couch potato. However, bones do much more than maintain your physical shape and help your body to do work. Bones not only partner with muscles to move body parts; they also help with new blood cell formation as well as act as a storage compartment for calcium.

Levers for Movement

If you have ever used a crowbar, you have demonstrated the concept of how levers work. Many bones in the skeletal system act as levers for the body, while the skeletal muscles (named such for their attachment to bones) provide the power for the work to be completed. Therefore, just like a lever system, the points where the skeletal muscles attach to the bones and which bones the muscles are attached to dictate the power and strength required to accomplish a task.

Skeletal muscles contract and pull in a straight line. One attachment is relatively fixed whereas the opposite attachment point is movable. The biceps muscle, for instance, has a relatively fixed attachment point (or origin of the muscle) at the scapula on the shoulder. At the opposite end of the biceps, the movable attachment point (or insertion of the muscle) is to the radius bone of the forearm. When the muscle contracts, it pulls on both ends, attempting to shorten the muscle toward the middle. However, because the shoulder is fixed, the only appreciable movement that can occur is for the forearm to be pulled closer toward the shoulder (flexing of the forearm). This action of the muscle, which is driven by the muscle contraction, is in fact a result of the involvement of the bones of the skeletal system.

Hematopoiesis

After just a few weeks following conception, the developing embryo (the unborn offspring from the time of fertilization to the end of the eighth week of gestation) will have grown too large for oxygen to diffuse to all the cells of the body. Red blood cells are formed from embryonic precursors and a primitive circulatory system is established. As the embryo grows and becomes a fetus (the unborn offspring from the ninth week through birth), the spleen and the liver serve as the location for new red blood cell formation. However, these organs serve other functions for the adult human. In adults, the blood cell production duties are contained in the long bones of the arms and legs.

Hematopoiesis is the production of blood cells from embryonic stem cells present in the bone marrow of the long bones. Bone marrow is the fatty substance in the bone cavity that helps produce blood cells. Erythropoiesis is the specific production of red blood cells, while leukopoiesis is the production of white blood cells.

Anatomy of a Word

erythropoietin

Erythropoietin (EPO) is a hormone produced by the pituitary gland that increases the rate of red blood cell production. This increase can be directly stimulated by living in an area that has a low atmospheric oxygen content, such as places with a high altitude.

Calcium Storage

Bone is principally composed of a hard, inorganic calcium phosphate matrix. While this gives bones the strength to resist gravity and support the movements of the body, it also gives the body a great reservoir of calcium. Unlike enamel, the calcium phosphate matrix of the teeth, bone is porous and filled with living cells that can repair bone.

Anatomy of a Word

osteoporosis

Osteoporosis is a disease of the elderly, particularly females, in which calcium and bone matrix is progressively lost. This eventually weakens the bones, making them more brittle and prone to breaking.

Just as sugar levels in the blood are regulated by two hormones (insulin decreases blood sugar and glucagon increases blood sugar), calcium is also closely monitored and regulated by two hormones. Parathyroid hormone functions to increase blood calcium levels while calcitonin decreases blood calcium levels.

AXIAL SKELETON

Never Grow a Wishbone Where Your Backbone Should Be

The body has more than 206 bones, all of which are grouped into two main categories. First are those bones aligned with the vertical plane or axis of the body. This is the axial skeleton and includes the skull, vertebral column, and rib cage. The other bones of your body, namely your limbs, make up the appendicular skeleton. Let's look at the axial skeleton first.

Skull

The skull, or cranium, is composed of numerous small and flattened bones that encase the brain, provide a base upon which the brain sits, and form an attachment point for connection to the vertebral column. The following are the different parts of the skull.

Neurocranium

The skullcap (membranous neurocranium, also called the calvaria) forms the roof of the skull. It is made up of several bones, which at birth are loosely joined together with spaces called fontanelles between them that allow for the rapid growth of the brain and later fuse into a single composite structure.

What is the "soft spot" on a baby's head?

The soft spot on the top of a newborn baby's head is actually the largest of several fontanelles (gaps into which the skull bones will later grow).

The skullcap is made up of left and right frontal bones (the forehead), right and left parietal bones (the top and most of the side), and an occipital bone (back lower portion of the skull). Around the area of the ears on the right and left are the temporal bones. These allow the passage of the auditory canal, seven cranial nerves, and major blood vessels.

Endocranium

This is the base on which the brain sits. Imagine the skullcap as the roof and the walls for the brain, and the endocranium as the floor. In human development, these bones first form as cartilage, and are later replaced by bone in the process called endochondral osteogenesis (covered in a later section). As with the skullcap, the base is a composite structure that grows and fuses into a single functional base for the brain.

Facial Skeleton

The final portions of the skull are the bones that make up the jaw and the face. Often referred to in anatomy class as the viscerocranium, it consists of the upper jaws (maxilla), lower jaw (mandible), and the bones of the nose and the palate of the mouth (nasal and palatine bones, respectively). Additional small bones are also present in the eye sockets and the deeper divisions of the nasal cavities.

Vertebral Column

This supporting column for the human body is a highly diverse and composite structure consisting of 33 vertebrae, small bones that interlock to protect the spine while allowing it to bend.

The most superior 7 vertebrae are called the cervical vertebrae and are often designated as C1–C7. Connecting the vertebral column to the occipital condyle, a protuberance at the base of the occipital bone, is the function of C1 (also called the atlas). The next vertebrae, or C2, is called the axis, and it allows the pivoting of the skull on the vertebral column. These are the only 2 named vertebrae in the body.

Anatomy of a Word

superior

Superior in anatomy does not refer to a snide manner, good quality, or excellent performance. It means "closest to the head." Inferior, then, means farthest from the head.

The next 12 vertebrae are the thoracic vertebrae (T1–T12). Each of these vertebrae projects at about a 45° angle; they form the connection points with the base of each rib.

Lumbar (lower back) vertebrae consist of the next 5 vertebrae (L1–L5), found in the lower portion of the abdominal area.

The remaining 9 vertebrae are actually fused into 2 units, the sacrum and the coccyx. The sacrum consists of 5 of the vertebrae (S1–S5) and functions as a portion of the pelvic girdle, the bones that connect the trunk and the legs. The last 4 small vertebrae are fused together into the structure known as the coccyx, or tailbone.

Rib Cage

Consisting of the ribs, interconnecting cartilages, and the bones of the sternum (breastbone), the rib cage is the protective structure surrounding the vital organs of the thoracic region including the heart and lungs. In addition, the rib cage helps you breathe.

Ribs

These 12 pairs of curved bones reach from the vertebral column and connect to the ventral (front) surface of the thorax (chest) via the sternum. The rib cage as a whole is capable of moving upward (and expanding) each time a person inhales. Likewise, when a person exhales, gravity pulls the rib cage downward (and inward) to its relaxed position.

The first 7 most superior ribs are referred to as the "true" ribs. While the composition of all ribs is identical, these first 7 are attached to the sternum via individual costal cartilages, tough pieces of connective tissue. The next 3 pairs are indirectly attached to the sternum via cartilages that join the costal cartilage of the seventh rib, and therefore do not have their own costal cartilage. The final 2 most inferior ribs (sometimes called "false" ribs) are called "floating ribs" because they are only attached to the vertebral column on the dorsal (back) side of the body, leaving the ventral (front) ends of the ribs unattached.

Sternum

The sternum is a composite flattened plate of bones that defines the ventral surface of the thoracic region of the human body—or, more simply put, your chest. The middle portion of the sternum, and the bulk of its mass, is the body of the sternum. For the superior portion, the manubrium is a broader square-shaped bone that is connected to the body of the sternum. The inferior portion of the sternum is formed by an arrow-shaped bone called the xiphoid process.

APPENDICULAR SKELETON AND JOINTS

Look, Ma! Hands!

While the bones of the axial skeleton define the vertical axis of the body, the appendicular skeleton makes up the arms and legs, as well as the girdles that secure the limbs to the axial skeleton.

Pectoral Girdle

The bones that support the arms and compose the pectoral girdle are the scapulae and the clavicles. Also known as the collarbones, the clavicles are slender bones that attach and anchor the scapula to the manubrium of the sternum. They also act as a platform for the attachment of muscles from the arms, chest, and back.

Anatomy of a Word

ligaments

Ligaments are connective tissues that attach bone to bone and resist stretching. Tendons are made from the same material as ligaments, but connect muscle to bone.

The scapulae (shoulder blades) are the broad and flattened bones clearly visible on the back. Each scapula projects from the shoulder toward the spine. The broad flattened surfaces of the scapula are locations for the attachments of large muscles of the back, including the supra- and infraspinous muscles as well as the subscapular muscle. The scapula is somewhat triangular in shape, with the base toward the spine and the point toward the shoulder. The surface of the point is actually a concave depression that serves as a location for the ball of the upper arm bone (humerus) to form a freely movable joint with the scapula. This connection is stabilized by the ligaments and tendons of the shoulder joint.

Arms and Hands

Human arms are made up of only 3 long bones: the humerus, radius, and ulna. The humerus is the single bone of the upper arm. The radius and ulna make up the forearm. The head of the humerus fits into the depression of the scapula called the glenoid cavity (glenoid fossa). At the opposite end of the humerus, the humerus forms a joint (the elbow) with the joining (articular) surfaces of the radius and ulna. The capitulum, a small knob on the end of the humerus, and the trochlea, a pulley-shaped part of the humerus, fit together with the head of the radius and the trochlear notch of the ulna, respectively.

At the ends of the forearms, a collection of bones cluster together into the wrists, hands, and fingers. The 8 small, irregularly shaped bones adjacent to the radius and ulna are collectively called the carpals, and form a base upon which the longer bones of the hands and fingers are attached to the bones of the forearm. The first long bones that extend from the carpals to the fingers are the metacarpals; they are attached to the 3 remaining bones of each finger, the phalanges.

Pelvic Girdle

The pelvis is a composite structure composed of the sacrum (along the dorsal, or back-side aspect) and 2 hip bones, the coxal, or innominate, bones. Together these bones create a bottomless, basket-shaped girdle that supports the lower abdominal organs as well as provides an attachment point for the legs to the axial skeleton. This open-ended structure in females is wider than in males to facilitate the passage of the baby through the birth canal.

The 2 hip bones are joined to the sacrum along the back of the body at the sacroiliac joint. This is a relatively fixed joint, especially when compared to the joining of the 2 pelvic bones along the front of the body at the pubic symphysis. During pregnancy and at delivery, this joint allows for the expansion of the pelvic girdle and delivery of the baby through the birth canal.

Each pelvic bone has 3 regions. The broad blades on the back of the pelvis are the ilium portions of the hip bones. The ventral portion of the hip bone is divided into a superior (pubis) and inferior (ischium), around which is a large opening called the obturator foramen. At approximately the location where all 3 portions of the hip bone connect is a large depression where the ball of the femur (leg bone) connects. This depression is called the acetabulum and forms the socket of the hip ball and socket joint.

Legs

The legs have a layout similar to that of the arms. They are composed of a single upper leg bone (the femur), 2 lower leg bones (the tibia and fibula), and bones of the ankles and feet. The femurs support the full weight of the body. They join with the pelvic bone via the head of each femur fitting into the sockets of the acetabula in the hip.

Anatomy of a Word

condyle

A condyle is a rounded protuberance at the end of a bone, generally where it attaches to another bone and forms a joint.

At the knee, the medial and lateral condyles of the femur form the articular (joining) surface with the smooth surface of the tibia. This lower leg bone supports the full weight of the body whereas the fibula actually joins with the side of the tibia rather than directly contributing to the knee itself. The remaining bone of the knee, which is actually suspended within its own tendon, is the patella (knee cap).

At the ankle, the lower portion of the fibula (lateral malleolus) projects outward on the outer side. This rounded bump is clearly prominent on the surface. It is here, at the surface of both the tibia and fibula, where the ankle joint is formed between the bones of the lower leg and the tarsal bones of the foot. The most prominent of these tarsal bones is the calcaneus (heel bone). Similar to the wrists and hands, the first long slender bones attaching the toes (phalanges) to the foot are the metatarsals.

Joints

While joints are often thought of as the movable points of the human body, some are actually completely immovable. Thus, the definition of a joint in strict anatomical terms is the connection between two or more opposing bones, which may or may not allow for movement.

Ball and Socket Joints

This type of joint, as seen in the hip and the shoulder, is produced when the ball, or head, of a long bone inserts itself into the bowl-shaped depression of another bone, either of the pectoral girdle (glenoid cavity of the scapula) or the pelvic girdle (acetabulum of the pelvic bone). This type of joint allows for a wide range of rotational movement and gives the most flexibility of the body's joints.

Synovial Joints

Synovial joints are also highly flexible, but only along a single plane. These are often called "hinge" joints, because they work like a door swinging open or shut. These joints (the elbow and knee) are covered in membranes that contain cells producing synovia, a lubricating fluid that can also cushion the surfaces of the bones that are in contact in much the same way that hydraulic fluid can withstand great pressures.

Fibrous Joints

These are immovable joints. In the adult human skull, they are seen as the suture lines between the bones of the skull. During the fetal period and into early childhood, the bones of the skull are not attached to one another in order to allow for the head to compress and pass through the birth canal. Later, these bones grow and fuse permanently with only the suture line as evidence that the bones were ever separated at all.

BONE GROWTH, REPAIR, AND DISEASES

It's a Blast!

Unlike the enamel of teeth, bone is living tissue, containing cells that can remodel the bone for normal growth and to repair any damage that may occur.

Bone Growth

During embryonic and fetal development, bones form in one of two ways, through intramembranous osteogenesis or endochondral osteogenesis.

The process of intramembranous osteogenesis forms the flat bones of the skull, clavicle, and sternum, among others. During embryonic development, connective tissue cells derived from the embryonic membranes start to gather and divide in the future sites of the bones. These cells form early bone cells called osteoblasts, which begin forming the calcium phosphate–rich bone matrix. This process continues until the bone cells completely surround themselves with bone matrix, at which time they become osteocytes and reside inside cavities in the bone called lacunae.

Since bone matrix is too dense for the diffusion of gases and nutrients, osteocytes are interconnected via cytoplasmic extensions that extend through small tunnels in the bone matrix and form a network of canals (canaliculi). Bone cells pass materials along the chain of interconnected cells much like a bucket brigade passes buckets of water to the fire. Each cell along the way uses the materials it requires to survive and passes the remainder along the chain to the next cell.

Thus, these bones resemble a cream-filled cookie with two outer layers of highly compacted bone called the tables with a region of more spongy bone (diploe) and a marrow cavity in between. The process of endochondral osteogenesis forms the long bones, such as the humerus and femur. They begin as small cartilaginous templates that exhibit the rudimentary shape of the adult bone. As the fetus grows, the cartilaginous templates also grow until blood vessels penetrate the middle of the long shaft (diaphysis). These vessels bring and deposit bone stem cells into the region of cartilage.

First, a collar of bone grows around the middle of the shaft. Since cartilage is fed by diffusion, this collar of dense bone suffocates and kills the cartilage cells, leaving room for the newly deposited bone cells. This primary ossification (bone forming) center begins the restructuring of the shaft of cartilage into bone. The outer portion of the shaft is composed of layers of compact bone with just enough room for the passage of blood vessels and nerves along the length of the bone.

In the center, the bone forms into shards called trabeculae, leaving much space and making the area appear spongy. This is the location of the marrow cavity, which fills with fat (yellow marrow) or blood stem cells (red marrow).

Vessels also enter the bulbous heads of the bones (the epiphyses), repeating the process of bringing bone cells to the site, killing the cartilage cells and forming a secondary ossification center. With bone forming from the middle of the shaft toward the ends and the secondary centers forming bone from the ends toward the middle, the only cartilage that remains is what is caught in between these two ossification centers at the junction between the epiphysis (the end part of a bone) and diaphysis (the midsection of the bone). These plates of cartilage persist until the individual is in his or her early to mid-twenties, at which time all cartilage will have been replaced with bone.

Bone Repair

Most people will experience a broken bone at some point in their life. In most cases, bone can repair itself because bone has a blood supply and is rich in living cells.

When a bone breaks, blood vessels within the bone are severed, and bleeding and blood clotting occur at the injured site. Shortly afterward, connective tissue cells that surround the bone in the layer called the periosteum begin to divide, migrate into the injured site, and lay down connective tissue materials, which are equivalent to the granulation tissue seen in wound healing of the skin. This material bridges the gap between the broken ends and stimulates changes that occur at both edges of the broken bone.

Within a few weeks, some of these connective tissue cells will change and first become cartilage cells. In the long bones, this is similar to how the bone first forms during embryonic development. This time, the cartilage provides structural material to fill in the gap between the broken ends where bone will be deposited. Osteoblasts then produce bone, which grows into the newly formed cartilage and completes the repair process. These new layers of bone are modified slowly over time, just as all bone is modified and reshaped as the body ages.

Bone Growth Diseases

As mentioned before, bones are rich in calcium, which is a key ion in many physiological activities in the body. Calcium can be deposited into bone as storage and also recruited from the bone to be used elsewhere. Over time, the balance between give and take of calcium from the bones can become unbalanced and result in weakening of the bone. Additionally, if certain key minerals and vitamins aren't a part of the diet, the process of normal bone growth can become unbalanced and lead to bone growing irregularly. The following are two such problems related to bone growth.

Osteoporosis

This degenerative bone disease occurs primarily in women of postmenopausal age and leads to a weakening of the bones in the body, sometimes resulting in bone fractures. Because the bone is losing mass or becoming more porous, the disease was named after this histological appearance. After menopause, hormone levels, especially those of estrogen, fall below what is needed to maintain the balance between bone absorption and bone deposition. Thus, cells called osteoclasts, which remove bone matrix and free calcium for return to the blood stream, become more active than the osteoblasts (bone-depositing cells).

Additionally, the hormones that control calcium reabsorption and storage also appear to change in favor of calcium restoration into the blood stream, which accelerates the loss of bone mass with age.

Treatment of osteoporosis

Hormone replacement therapy (HRT) is a clinical option for these patients; however, certain hormones have been shown to increase the risk of breast cancer, and HRT must be done under the close supervision of a physician.

Rickets

Bones in individuals with rickets are typically seen as bowing in the long bones of the legs or otherwise misshaped bones within the arms or legs. The underlying cause of this disease is inadequate calcium deposition into bones, resulting in thinner and weaker bones (which bow under the weight of the body). While calcium may be present in the diet, what is missing is vitamin D. This is essential for proper absorption of calcium across the wall of the intestinal tract and transport into the blood stream where it can be utilized by the body. Likewise, calcium deficiency may also lead to rickets; however, this typically only occurs in areas where people, especially children, are living in conditions of famine and starvation.

MAJOR SKELETAL MUSCLES

You've Got to Move It Move It

Skeletal muscles are one of three types of muscles in the human body (the other two are cardiac and smooth). They are the engines that enable the human body to move and perform physical tasks as simple as holding a coffee cup or as complex as ballet dancing. Depending on how you divide up the skeletal muscles in the human body, there are well over 600 named muscles, with some estimates reaching more than 800.

Muscles of the Head and Neck

The lips of the human body, which can be so very expressive, are in fact a single oval-shaped muscle called the orbicularis oris. Likewise, surrounding each eye is a circular muscle called the orbicularis oculi, which is responsible for the movement of the eyelids. There are many other smaller muscles for facial expressions, including the frontalis, the forehead muscle that raises the eyebrows, and the buccinators, which form the cheeks and pull them inward when contracted. For chewing, no other muscle is as important as the masseter. Its insertion into the mandible leads to the powerful closing of the jaw during eating.

Several muscles of the neck help to move the head from side to side and up and down. The most prominent of these muscles is the sternocleidomastoid. It runs from the temporal side of the skull around to the ventral (front) side of the lower neck and connects with the sternum, which leads to the action of pulling the head to the side of the contracted muscle. Since this is a paired muscle, with one on the right and one on the left, they give the appearance of a V-neckline on the ventral lower neck.

On the dorsal (back-side) surface, a portion of a larger muscle, the trapezius, connects to the back of the skull as well as the cervical and thoracic vertebrae in a fanlike pattern and inserts into the scapula, allowing various movements of the scapula.

Muscles of the Chest and Shoulders

The muscle that forms the bulk of the human chest is the pectoralis major, often simply referred to as the pecs. Much like the trapezius, the pectoralis attaches to the sternum in a fan-shaped pattern, with the fibers coming to a point and inserting into the upper portion of the humerus of the forearm. Contraction of this muscle causes the humerus to be pulled toward the ventral midline of the body, therefore adducting the arm.

Anatomy of a Word
adduction and abduction

Adduction is a muscular action in which a body part is moved toward the midline of the body. You can remember this by saying the movement *adds* to the midline. If the body part is removed from the midline, the action is referred to as abduction (which means "taking away").

Abduction of the arm is largely accomplished by the action of the deltoid muscle. This large muscle composes the bulk of the shoulder. It has dorsal, ventral, and medial portions attached to the scapula, clavicle, and humerus. Contraction of this muscle raises the arm. While the pectoralis and the deltoids are the most prominent of the muscles that move the arm, many other smaller and deeper muscles mediate all other actions of the arm and allow it a wide range of motion and action.

Muscles of the Arms

In the upper arm, two muscles work antagonistically, or against each other, in flexing (bending) or extending (straightening) the forearm. The large muscle on the ventral surface is called the biceps brachii, which is incompletely divided into a long and short head (hence the prefix *bi-*, meaning "two"). The biceps pulls the forearm closer to the upper arm when contracted and thus flexes the arm. Straightening the forearm is the task of the triceps brachii on the ventral side of the upper arm. The name of this muscle stems from its division into three heads, two of which are visible superficially with the third being deep.

The forearm has a highly intricate collection of small muscles that assist in moving it as well as moving the wrists and fingers. The muscles compose the most proximal portion of the forearm and connect to bones of the wrists and fingers via long tendons, which are used to perform specific actions. Flexors contract and extensors straighten.

Anatomy of a Word

proximal

Proximal generally means "closest to the body." In some instances it is used to refer to the area closest to the point of attachment.

Additionally, other muscles mediate pronation (twisting of the hands from a palms-up position to a palms-down position). Conversely, other muscles called supinators rotate the hands back to a palms-up position.

Muscles of the Back and Hips

Starting at the highest aspect of the back (dorsal surface) is the aforementioned trapezius muscle. The remainder of the superficial back muscle is the large fan-shaped latissimus dorsi. Attached along the vertebral column from the thoracic region downward to the sacrum, it extends to a point under the arm and inserts into the humerus. In bodybuilders, contraction of this muscle causes the edge of the muscle to extend out laterally on both sides of the body, much like the hood of a cobra fanning out around the head.

Other deeper muscles of the back include the rhomboids (major and minor), which also attach to the vertebral column and cause various movements of the scapula when contracted.

Making up the dorsal hips (also called rear, posterior, and buttocks), are the three gluteus muscles. The gluteus maximus is the largest of the three and produces the bulk of the posterior hip tissue. When contracted, this muscle extends the hip and brings the thigh into a straight line with the hip. Along the side of the hip near the waist is the gluteus medius, which along with the deep gluteus minimus abducts and rotates the thigh to the side.

Muscles of the Abdomen

While not everyone has a clearly visible 6-pack, everyone does possess the paired rows of five muscle bundles that lie along the ventral midline of the body, called the abdominal muscles, or rectus abdominis. It is the middle three muscle bundles that are often called the 6-pack. Along with the rectus abdominis, the following muscles are responsible for tensing the abdominal wall and compression of the abdominal contents:

- From the rectus abdominis running along the side of the body is the external oblique muscle.
- Just beneath the external oblique is the internal oblique.
- The transversus abdominis, the deepest of the abdominal muscles, lies underneath the internal oblique muscle.

Muscles of the Legs

The ventral surface of the thigh is made up of a major group of muscles collectively referred to as the quadriceps (quadriceps femoris). This group comprises the rectus femoris, vastus lateralis, vastus medialis, and vastus intermedius. The prominent rectus femoris is the middle and superficial muscle of the group, with the vastus lateralis and medialis on the outside and inside of the thigh, respectively. The intermedius is a deep muscle lying underneath the rectus femoris. Together, these muscles function to extend the knee (straighten the leg).

On the back of the thigh (dorsal side), the group of muscles that function to flex (bend) the knee and pull the heel upward toward the hip is called the hamstring group. The two superficial muscles of this group are the biceps femoris and the semitendinosus. These paired muscles lie along the midline of the dorsal thigh, with a deeper, more medial (inside) muscle called the semimembranosus completing the group.

The calf muscle is formed from the two-headed gastrocnemius muscle. This muscle is connected to the heel (calcaneus bone) via the long, tough calcaneal (Achilles) tendon. Contraction of this muscle pulls the heel upward and straightens the foot. Imagine someone standing flat-footed and then rising up on the tips of her toes, like a ballet dancer. As in the hand, many smaller muscles function to flex, extend, and rotate the foot in several different directions.

NEUROMUSCULAR JUNCTION

The Function of the Junction

The nervous system instructs muscles to contract through electrical signals that flow from neuron to neuron. However, no physical contact exists between neurons and target tissues such as muscle. Therefore, the electrical signal of neurons must be turned into a chemical signal that can diffuse across the space (called the synaptic cleft) between the neuron membrane and that of the muscle. Chemical receptors on the muscle cell detect the chemical and activate a signaling cascade that will lead to the regeneration of the electrical signal within the muscle cell. This regenerated electrical signal causes the contraction. In the following section, each component of this junction, called a synapse, and signal transduction mechanisms are discussed.

Anatomy of a Word

signal transduction

Signal transduction is a biological process in which a cell converts one type of impulse or signal into another. For example, a neuron creates an electrical impulse that is converted to a chemical by neurotransmitters, then transferred to a muscle cell that converts the chemical back into an electrical impulse.

Motor Nerve

Signals from the nerves result in the contraction of skeletal muscles. In neurons, differences in ions between the inside and the outside of the cells result in a membrane voltage, with one side of the plasma membrane being more or less positive than the other. Think of this as being similar to the way a battery in your remote control has a positive and negative end (pole).

As the neuron receives a stimulus, membrane channels allow ions to move across the membrane and cause the voltage of the membrane to change. This localized change of the membrane influences neighboring membrane channels, and also opens channels in a wavelike progression that flows along the axon of the neuron toward its target. When this moving change of voltage (action potential) reaches the end of the nerve, neurotransmitters stored in vesicles fuse with the terminal (presynaptic) membrane and are released into the space between the neuron membrane and the target cell.

Think of this process as similar to what happens when speech is converted to sign language. The medium is different but the message is the same. A thought or sentence can be spoken aloud, converted to sign language and signed (communicating the same information), then converted back to spoken language and spoken aloud.

For motor neurons that stimulate skeletal muscle contraction, the neurotransmitter secreted is acetylcholine. Thus, the electrical signal of the nerve has been transduced into a chemical signal that can diffuse across space.

Skeletal Muscle Membrane

To regenerate the electrical signal in the muscle cells, neurotransmitter receptors are localized in close proximity to the neuron terminal on the muscle membrane. In skeletal muscle, these acetylcholine receptors are named nicotinic receptors. When 2 molecules of acetylcholine bind to the receptor, sodium and potassium ions are allowed to flow into and out of the cell through the nicotinic receptor. This localized change in voltage also leads to changes in the membrane farther along the muscle surface. These regions have membrane channels like those along neurons, which facilitate a wavelike action. Once activated by the nicotinic receptors, the action potential is regenerated and flows over all of the muscle membrane. Because the plasma membrane of skeletal muscles invaginates—folds into a sac or cavity—into the muscle cell, forming tunnels called T-tubules, the action potential spreads rapidly and completely throughout the muscle cell and results in a muscle contraction.

Motor Units

For muscles to contract, energy (in the form of ATP) must be expended. To conserve energy and use only as many muscle cells as are required to accomplish a task, muscles are divided up into functional units called motor units, which consist of a single motor neuron and all of the muscle cells that the neuron is connected to and controls. To initiate a contraction, the central nervous system activates only a few motor units and then progressively activates more and more until the work is accomplished.

What is an example of a motor unit?

For a muscle whose primary function is strength (quadriceps group), there may be several hundred muscle cells connected to a single motor neuron. However, for muscles that require less strength and more control, such as the ciliary muscle that controls the shape of the lens in the eye, only a few muscle cells are attached per neuron.

MUSCLE CONTRACTION

How the Work Is Done

Muscles have only one function, and that is to contract. For skeletal muscle, this results in pulling both ends of the muscle (and the structures to which they are attached) closer together. This is mediated by the two contractile proteins, actin and myosin, and the sliding movement that occurs between them during a contraction.

Banding Pattern of Striated Muscle

Skeletal muscle (and cardiac muscle) is often described as striated muscle because of the striped appearance of the individual muscles cells when observed through a microscope. The light and dark bands are caused by the amount of light that may pass through a particular region of the muscle. The denser areas are darker, leaving the less dense regions lighter by comparison. These darker bands, called A bands or anisotropic bands, contain myosin filaments. Myosin is a type of protein; groups of the protein together are called myosin filaments and have a threadlike appearance. These proteins, shaped much like the human arm, are wrapped around the cylinder of the myosin filament and secured by a light chain, leaving the remainder of the myosin molecule (heavy chain) free to move. While this creates a high-density area of the muscle, each end of the dark band also contains actin molecules that insert themselves between and overlap the myosin filaments, resulting in the highest density area and the darkest part of the A band.

However, these actin filaments do not extend to the center of the myosin filaments. They only insert themselves about a quarter of the length of myosin on each end, leaving the middle portion of the

A band composed of only myosin. This region is therefore less dense than the ends and is seen as a light region in the center of the A band called the H band. Additionally, in the very center of the A band, and also in the center of the H band, is a dark line consisting of structural molecules that assist in holding the myosin filaments in the proper position. This dark line is the M line and also marks the center of the contractile unit of skeletal muscle called the sarcomere.

On either side of the dark band are lighter regions called the isotropic (I) bands. These areas contain only actin filaments, which are much less dense than myosin. Each actin filament is composed of two strands of filamentous (F) actin that are twisted together. Each F actin strand is formed from polymers of G-actin or globular actin molecules. This gives the F actin the appearance of a pearl necklace, in which each pearl represents a G-actin molecule. The I band is interrupted by a dark line (Z line or Z disk) that defines the center of the I band and is composed of structural molecules much like those of the M line in the dark band. These molecules also assist in maintaining the proper spacing of the actin molecules, which is critical to the sliding filament action that occurs during a muscle contraction.

Anatomy of a Word

sarcomere

A sarcomere is the basic segment of a muscle, a structural and functional unit that aids in the contraction of the muscle. It contains an entire A band and two halves of I bands on each end. During a contraction, the actin molecules on the ends of the cell are pulled toward the M line. The Z lines are also pulled closer together and the sarcomere as a whole shortens.

Accessory Proteins

While the structural molecules of the M and Z lines are important for the alignment of the thin actin and thick myosin filaments, other molecules play essential roles in helping regulate a muscle contraction and in returning the muscle to the relaxed state. One such molecule is tropomyosin. Two of these filamentous molecules run along the grooves between the two F actin molecules (which creates a thin actin filament) and function to mask sites on each G action molecule where myosin can bind. When tropomyosin is in this position, the muscle is relaxed.

Attached to tropomyosin is a multiunit molecule called troponin. One of troponin's three subunits, troponin I, binds to a region of the actin molecule. The troponin T subunit binds to the tropomyosin molecule. The final subunit, troponin C, is capable of binding to a calcium ion. Thus, calcium unmasks the myosin bound regions of the actin molecules and leads to a muscle contraction.

Calcium and Its Role

In the relaxed state, calcium is stored in muscle cells inside organelles called sarcoplasmic reticula. This calcium reservoir is responsible for releasing calcium upon nerve stimulation, but also for pumping calcium back inside when the nerve signal ceases, signaling the muscle to relax.

Voltage-gated calcium release channels are closely associated with the T-tubules described earlier. As the action potential spreads across the surface and down into the T-tubules, it leads to the rapid release of calcium from the sarcoplasmic reticulum and its rapid spread throughout the cytoplasm of the muscle cell where it begins a muscle contraction through contact with the accessory proteins.

Sliding Filament Motion

When the head of myosin connects with a G-actin, a cross bridge is formed. This is only made possible because prior to the connection, a molecule of ATP was bound inside the myosin head and split into separate molecules of ADP and phosphate, which remain inside. Once the cross bridge is formed, however, the phosphate molecule is released, which triggers a change in the myosin molecule called a power stroke. Since the myosin head is bound to actin, the power stroke is the action that results in the sliding of actin closer to the M line. Since this is happening on either side of the M line, each Z line is moved closer together and the muscle as a whole contracts.

This single contraction cycle only provides a fraction of the shortening distance for the muscles. Several repeated cycles of contraction will need to be accomplished for the entire muscle to shorten the full distance. Thus, each myosin head will need to undergo a power stroke, release, reset, and repeat a number of times to shorten the muscle fully. Consider a tug-of-war team. Each individual pulls on the rope trying to pull the other team across a line. During the contest, it will be necessary for members to release their grip and pull on the rope from a new position. If each member released at the same time, the team would lose. So the members release and form new grips in an alternating fashion. Such is the case for cross bridges for striated muscle.

What causes rigor mortis?

Rigor mortis, or the "stiffness of death," occurs as skeletal muscle cells contract at the time of death from an explosion of electrical signaling throughout the body. This signal doesn't cease and the muscles remain locked in continuing cycles of contraction. However, since ATP is required for contraction and for resetting, the cycles stop when the supply of ATP is exhausted, at which time the muscles are locked in the last contracted power stroke and unable to rest.

During the release and resetting period, ATP is hydrolyzed into ADP and phosphate, and as long as the myosin binding site is still exposed on actin, a new cross bridge and power stroke can occur.

At the end of a muscle contraction, the nerve stops sending neurotransmitters and stops the action potential on the muscle cell. This halts the release of calcium and initiates the active transporting of calcium back into the sarcoplasmic reticulum. With calcium no longer available to bind to the accessory protein troponin, tropomyosin slides back into its original position. This blocks the binding sites on actin and leaves myosin in its resting position.

Energy Sources for Contraction

Energy, in the form of ATP, is required for muscle to shorten. At the onset of muscle contraction, the raw material used to produce new ATP is pulled from the plasma of the blood and becomes circulating glucose, which can be used for the immediate production of energy. As the work continues or if the work intensifies and as plasma glucose is in short supply, muscle cells draw on plasma triglyceride resources, such as those present in fat cells. For high-intensity work, the demand for energy by muscle cells exceeds what can be supplied by plasma molecules or fat cells. In this case, the body recruits glucose from storage. The liver stores glucose as glycogen, which can be tapped to return glucose into the plasma, supplying the muscles with a rich source of raw material for immediate ATP production.

MUSCLE DISEASES AND DISORDERS

Conditions That Affect the Muscles

Problems with the muscular system can range from mild and temporary discomfort to the extreme of muscle loss (atrophy) and death. Most will experience the first at some point, while the extreme conditions are much less common. However, it is likely that you will encounter someone who has been affected by muscular disease in your lifetime. Infections, autoimmune diseases, and cancer can affect the muscles.

Injury and Overuse

If you've ever strained your back moving furniture, then you know that injuries to muscles are a common problem. Some injuries that appear to be muscle-related actually affect the connective tissue that attaches a muscle to a bone (a tendon) or a bone to a bone (a ligament). Here are two common injuries and their causes:

- **Sprains:** A *sprain* affects the ligament—it is stretched or torn, causing swelling, tenderness in the affected area, and pain. This commonly occurs when you fall or accidentally twist a joint.
- **Strains:** A *strain* affects a tendon or muscle—as with a sprain, the tendon or muscle is stretched or torn, causing swelling, tenderness, and pain. With a strain you might also get muscle spasms and difficulty moving the muscle. Sports injuries are quite commonly strains. Strains can happen because of overuse over a period of time or can occur suddenly.

Rest, ice, and compression are the usual treatments for both sprains and strains.

Muscle Spasms

Resulting largely from dehydration, muscle spasms are involuntary and repetitive or constant contraction of a muscle without being able to relax. Often uncomfortable or painful, these spasmodic contractions occur because the muscle has received an inappropriate electrical signal to contract and remain contracted until the electrical signal stops. Recall that muscles are signaled to contract by a nerve stimulating a change in voltage on the muscle membrane. If the concentration of these charged ions changes because of dehydration and become more concentrated around the muscle cell, this can yield the same change in voltage as a nervous stimulus and lead to a contraction. Since this is not regulated as a normal contraction, the signal doesn't stop until the ions become balanced once again.

Often, therapists or trainers massage and stretch the affected muscle to increase the circulation in and around the muscle in order to balance the ions as quickly as possible. Additionally, immediate rehydration will aid in preventing the recurrence of the cramps.

What is the difference between a muscle cramp and a spasm?

Muscle spasms are involuntary contractions of the muscle, and a cramp is what we call the result of the spasm, so they are basically the same thing. A muscle spasm can create other symptoms associated with a cramp (such as pain). Occasionally medications can cause muscle spasms.

Muscular Dystrophy

The most common form of muscular dystrophy is Duchenne muscular dystrophy (DMD), which occurs most commonly in childhood. This debilitating disease not only reduces muscle mass and mobility; it also considerably shortens life expectancy.

The cause of DMD is a mutation of the gene for dystrophin, a protein that interconnects the muscle cytoskeleton (the intercellular proteins) to the extracellular environment through the muscle membrane. Loss of dystrophin function results in highly disorganized muscle, smaller muscle mass, and an increase in connective and inflammatory tissue.

Becker muscular dystrophy is a variant of DMD where dystrophin is shortened but remains functional, and is therefore a less severe form of DMD.

Are there other causes of muscle problems?

Some diseases that affect the muscles are actually neurological diseases. Parkinson's disease is one example. It causes muscle tremors, slowness in movement, and decline in agility, but the cause is related to the death or break-down of neurons, not muscle problems. This is why it is classified as a neurological disorder. Other neuromuscular diseases include multiple sclerosis, amyotrophic lateral sclerosis (ALS or Lou Gehrig's disease), and myasthenia gravis.

NERVOUS SYSTEM SIGNAL TRANSDUCTION AND NEUROTRANSMITTERS

Intracellular Communication 101

The nervous system is one of the major control centers of the body. It is composed of sensory receptors and nerve tracts, which bring information about the body (internal and external) to the brain. The brain processes the information and determines how to respond. Those responses leave the brain via different nerve tracts, travel through the spinal cord, and are dispersed through nerves to the appropriate target, such as different tissues. All of this information is moved around the body via ions and chemicals in and around the nerve cells. Therefore, the signal transduction of information is essential for the nervous system.

Signal Transduction

As discussed earlier, neurons use their plasma membrane to selectively concentrate ions in either the cytoplasm or outside of the cell, which affects the voltage of the membrane. When cells are inactive (at rest), the metabolic machinery of the cell works to create an appropriate concentration of two critical ions: sodium and potassium. A membrane protein called the sodium-potassium pump (Na/K pump) uses energy to move sodium outside of the cell, while at the same time moving potassium inside. This, as well as the number of fixed charged particles on the inside of the cell (such as DNA and charged proteins of the cytoskeleton), results in the interior of the cell being more positive than the outside of the cell, creating a membrane voltage of approximately -70 mV (a millivolt, mV, is one-thousandth of a volt). Since the cell is at rest, this voltage is termed the resting membrane potential.

Voltage-Gated Receptors

When stimulated, protein receptors cause some ions to diffuse out of their compartment and alter the membrane voltage. If this change is of sufficient magnitude, some membrane proteins actually change their shape. These shape-changing proteins are said to be "voltage-gated" channels. For instance, one such channel will change shape at a voltage of -56 mV in such a way as to allow sodium ions to diffuse into the cell. Since sodium is positively charged, this causes the membrane voltage to become more positive and possibly affect other voltage-gated channels that respond at different voltages. Potassium voltage-gated channels open when the membrane potential is in the positive range.

Action Potential

The process of channels opening and closing and ions changing place creates a wave of voltage changes that can affect nearby areas of the membrane, causing them to undergo the same changes. This wave of voltage changes moves along the axon of a nerve cell in much the same way that the wave moves in the stands at a football game. In the stadium, as a person stands up then sits, the next person knows to then stand and sit, and so on.

On the cell membrane, sodium channels open at -56 mV and the movement of sodium causes the membrane potential to become more positive (and move toward 0 mV). This change is called depolarization of the membrane. Because sodium flows into the cell more rapidly than needed, the sodium channel inactivates quickly after it opens to prevent too much sodium from coming in. However, this still allows enough sodium in to change the membrane potential into the +30 mV range. Inactivation is different from the channel being closed, which occurs more slowly. Inactivation means that although the shape of the channel remains the same, another part of the protein has moved into place to block the passage of any further ions until the entire protein can reset and close.

As sodium rushes into the cell, it disperses throughout the cell, leading to an increase of the membrane potential in the area of neighboring sodium voltage-gated channels, which are still closed. This leads to the neighboring sodium channels opening if the membrane potential in that area also reaches -56 mV. This will continue the length of the axon as long as there are sodium voltage-gated channels to be opened.

When the membrane potential enters the positive realm, potassium voltage-gated channels open and potassium rushes out of the cell by diffusion. Since potassium is positive, the potential becomes more negative, therefore repolarizing the membrane. As with the sodium channels, the potassium channels open and inactivate quickly, but not fast enough to prevent the potential from becoming even more negative than the resting membrane potential of -70 mV. This period is termed hyperpolarization, and although the voltage has returned to that near resting potential, the concentration of sodium and potassium ions has been reduced. To reset the ion concentrations, the Na/K pump shifts into high gear to pump sodium out and potassium in to re-create the high concentration gradients sufficient to have another action potential occur in the area of the cell.

Neurotransmitters

Action potential works very well to transmit electrical signals within a cell. However, these electrical signals cannot move across space and continue in an adjacent cell spontaneously. Thus, to functionally interconnect neurons with each other and to their target tissues, the electrical signals are transduced (changed) into chemical messengers that can be secreted, diffused through space, and be detected by the receiving cell that can transduce the now chemical signal back into an electrical action potential.

Types of neurotransmitters

Neurotransmitters are the chemical messengers of the nervous system. Several different types of these molecules function with different tissues, are secreted by specific neurons, and elicit prescribed effects from the target tissues:

- Acetylcholine (ACh) is the neurotransmitter most commonly secreted from motor neurons. Because of the interaction with ACh and its receptor, the result is always an increase in the membrane potential of the receiving cell. Therefore, ACh is said to always be excitatory in nature.
- Norepinephrine affects smooth and cardiac muscle as well as glands of the body. Norepinephrine belongs to a group of neurotransmitters called catecholamines (molecules derived from the amino acid tyrosine), and functions largely in the involuntary nervous system to control those body functions either during rest or in fight-or-flight mode.
- Other neurotransmitters, such as dopamine, serotonin, and gamma amino butyric acid (GABA), are found in the brain and control hunger, behavior, mood, and overall brain activity.

Chemically Gated Receptors

Without a receptor to detect and instruct the cell, neurotransmitters would have no effect. Thus, the receptors are equally as important as the neurotransmitters themselves.

On skeletal muscle cells, the ACh receptor (nicotinic ACh receptor) binds to 2 molecules of ACh, changing the shape of the receptor. This opens a channel in the protein through which sodium and potassium can diffuse, resulting in a positive change in the membrane potential. Since these are chemically gated responses, the process is not referred to as a depolarization; rather, it is called an excitatory potential.

Smooth and cardiac muscles have a different ACh receptor, the muscarinic type. Binding of a single ACh molecule leads to a shape change in this protein, but it is not an ion channel as with the nicotinic receptor. The shape change leads to a signaling cascade of molecules, much like a series of dominoes knocking each other over, that ends in either opening or closing the ion channels. Depending on the ion and the directional effect, this type of signaling is either excitatory or inhibitory (decreasing the membrane potential).

Receptors for other neurotransmitters in other tissues similarly transduce signals and lead to alterations in their membrane potentials and yield an appropriate physiological response.

Synapse

When the end of a nerve cell (the axon terminal) approaches the target tissue, such as a skeletal muscle cell, a space is created between the membranes of each cell (the synaptic cleft). When released from the axon terminal, neurotransmitters diffuse across this space. Being on the receiving end of the chemical messages, the skeletal muscle cell membrane and its receptors are said to be part of the post-synaptic cell or membrane. As previously described, voltage-gated channels open and the action potential is regenerated in the new cell using the same machinery.

BRAIN AND SPINAL CORD

Your Body's Central Processing Unit

The major component of the central nervous system (CNS), the brain comprises the cerebrum, cerebellum, and the brain stem. All activities of the body are controlled from the brain, whether that is sensing the internal and external world, reflexively or consciously responding to stimuli, coordinating all body movements, or simply dreaming of a better world.

The spinal cord is the extension of the CNS throughout the axial core of the body. The spinal cord receives and sends information and transmits that information to and from the brain.

Cerebrum

The cerebrum, or the forebrain, is the organ used for conscious thought, for receiving and perceiving sensory information, and for initiating motor responses. On its surface are numerous folds (gyri) and grooves (sulci) that allow for the outer cortex, a layer of neural tissue, to be packed into a smaller space.

Each hemisphere of the cerebrum is functionally divided into regions. These regions of the brain, called lobes, are responsible for specific functions:

- The frontal lobe is involved in decision-making, overriding instinctual urges that may not be appropriate socially, and planning for future events. Long-term memories are also formed in this region of the brain. This lobe makes up the front half of the cerebrum.

- The parietal lobe connects to the frontal lobe at the midpoint of the brain. It is the location of the somatosensory center of the brain that processes sensory information such as touch and is a major integrating center for incoming information.
- On the side of the cerebrum is the temporal lobe. This region is responsible for the processing of vast amounts of information related to incoming sound and vision. New memories as well as understanding spoken language occur here.
- The final lobe, the occipital lobe, is the very back portion of the brain and functions as the primary visual center of the brain to interpret, integrate, and perceive visual information.

Cerebellum

Positioned beneath the occipital lobes of the cerebrum, the cerebellum functions principally for the coordination of motor activity. It consists of two major lobes, called cerebral hemispheres, with a smaller interconnecting region, the vermis. New research is shedding light on the other activities that the cerebellum may be involved in, such as learning, mood, and behavior.

Brain Stem

The posterior portion of the brain, which is connected to the spinal cord, is termed the brain stem and is composed of the pons and the medulla oblongata. The pons is the enlarged beginning of the brainstem, and is located where nerve tracts that enter and exit the cerebrum are organized. Additionally, the pons has regulatory centers for breathing and heart rate. Extending downward and comprising the nerve tracts to and from the spinal cord is the medulla oblongata. This region also serves as a critical regulator of basic body functions such as respiration and heart rate. The brain stem continues to function in individuals who have clinically been determined to be "brain dead."

Spinal Cord Gray Matter

White matter makes up the outer layers of the spinal cord, while the gray matter (neural cell bodies) is in the center of the cord. The spinal cord has a gray matter connection called the commissure. In each half of the cord, the remainder of the gray matter is present as the dorsal horn and the ventral horn. The dorsal horn contains synapses between sensory neurons (bringing in information from the periphery) and interconnecting neurons that transmit the signal elsewhere. The bodies of the interneurons are located here. Cell bodies of the motor neurons (sending information out to the periphery) are located in the ventral horn of the gray matter.

Anatomy of a Word

interneuron

An interneuron is a type of neuron that links sensory neurons with motor neurons. They exist only within the central nervous system.

Spinal Cord White Matter

The white matter surrounding the gray matter of the spinal cord consists of the myelinated axons of nerves that carry sensory information to the brain (ascending or afferent fibers) or motor information from the brain (descending or efferent fibers), down the cord to go out to targets in the body. These nerves form synapses with nerve cell bodies in the gray matter of the spinal cord before proceeding either up or out of the cord.

PERIPHERAL NERVOUS SYSTEM

The Other Nervous System

The nerves present outside of the central nervous system are collectively termed the peripheral nervous system (PNS). Most emanate from the spinal cord and carry sensory and motor information to and from the CNS and are called spinal nerves. However, others come from the brain or brain stem as cranial nerves to service the needs of the body.

Cranial Nerves

The human body contains 12 cranial nerves, numbered 1–12 (usually with roman numerals, I–XII). The first 9 nerves function with all of the special senses or control movement of the face and eyes. Cranial nerve X (10), the vagus nerve, serves functions and organs throughout the body and is the most widespread of the cranial nerves. Nerves XI (11) and XII (12) control head turning and tongue movement, respectively.

Spinal Nerves

Composed of sensory and motor fibers, spinal nerves travel laterally from the spinal cord and course throughout the body. They have two roots, a posterior (or dorsal) and anterior (or ventral) root, that come directly from the spinal cord. The dorsal root carries sensory information and ends at a collection of cell bodies called ganglia. The ventral root carries motor information. The posterior and anterior roots meet at the distal end.

From the dorsal root ganglia, sensory cells send axons into the dorsal horn to form synapses with interneurons of the spinal cord. Motor nerve cells extend their axons away from the spinal cord via the ventral root.

AUTONOMIC NERVOUS SYSTEM

Nice Reflexes

The CNS is organized into separate control centers for sensory information and motor output. The motor side is divided into the somatic portion (control for skeletal muscles) and the autonomic (subconscious control of smooth and cardiac muscle as well as glands). This division is further subdivided based on anatomy, neurotransmitters used, and physiological effects that result.

Autonomic Reflex

Unlike the knee-jerk reflex, which consists of one sensory nerve and one motor nerve for an immediate and subconscious reaction, the autonomic reflex involves a series of two motor neurons that deliver motor instructions to target tissues. The first motor neuron leaves the ventral horn of the spinal cord as all other nerves do; however, before reaching a target, it forms a synapse with a second neuron in the periphery of the body, typically in a collection of nerve cell bodies (ganglia). These two nerve cells are named relative to the ganglion: the preganglionic nerve and the postganglionic nerve. The postganglionic nerve innervates the target tissue and elicits the response.

Sympathetic System

Also known as the fight-or-flight nervous system, the sympathetic system instantly prepares the body for intense physical activity. The effects of this system are facilitated and initiated by neurotransmitters and a related hormone, epinephrine (adrenaline). Via the blood stream, this hormone can quickly spread throughout the body and cause the entire body to prepare for action.

Preganglionic sympathetic nerves leave the spinal cord in the middle of the spinal cord from the first thoracic vertebrae (T1) to the second lumbar vertebrae (L2). These fibers are said to form in the thoracolumbar region.

As soon as the preganglionic nerves leave the spinal cord and become part of spinal nerves, they exit the spinal nerve via a pathway called a white ramus. This is not unlike a car exiting an interstate highway to gain access to a side road. In this case, the nerve exits the spinal nerve to gain access to one of the ganglia that form an interconnected chain that mirrors the spinal cord on the right and left side. This sympathetic chain of ganglia allows dispersal of nervous system information along a wide stretch of the thoracic and abdominal region, and controls activities such as shutting down much of the digestive system during a sympathetic response.

From the synapse in the sympathetic ganglia, the postganglionic nerves enter the spinal nerve via a gray ramus (the on ramp) and run to and innervate the target tissue. However, some nerves bypass the chain of ganglia and form two large nerve bundles, the greater and lesser splanchnic nerves. These course through the body and form synapses with postganglionic nerves in collateral ganglia, such as the celiac, superior mesenteric, and inferior mesenteric ganglia, before extending to the target tissue.

Effects

The physiological effects during a sympathetic response rev up the body in preparation for intense activity. These include:

- increased heart and respiratory rate
- dilation of the pupils
- increased blood pressure
- direction of more of the body's blood supply to the skeletal muscles

These changes occur partly from dilation of blood vessels to the muscles; however, much of this shift is because of the constriction of blood to the digestive and urinary system. Those systems deemed nonessential for immediate survival are restricted in blood supply (to a minimal level) to divert those resources to the organs and tissue needed for short-term survival.

Parasympathetic System

The opposite of the sympathetic system, the parasympathetic system is in control during those periods of rest where the body can perform nonactive functions such as digestion. Thus, this system is often referred to as the rest-and-digest system. Glandular secretions of the GI tract, peristaltic movement, and absorption in the lower alimentary canal occur. Heart and respiratory rates decrease, as does the amount of blood going to skeletal muscles.

The preganglionic fibers exit the spinal cord in the areas not used by the sympathetic system: the cervical and the sacral regions. Sympathetic nerves are in the middle, and the parasympathetic nerves are at the top and bottom of the spinal cord. In the head region, cranial nerves serve as preganglionic neurons for the parasympathetic system. These run practically the entire distance to the target before forming the synapse with postganglionic neurons in what are termed terminal ganglia. Other nerves terminate within the organ itself.

Neurotransmitters and nerves

Nerves use different neurotransmitters to transduce their signals:

- Since the preganglionic nerve always excites the postganglionic nerve, the neurotransmitter used by all preganglionic nerves (to signal the postganglionic nerve) is acetylcholine (in both sympathetic and parasympathetic systems).

- Norepinephrine is the neurotransmitter used by sympathetic postganglionic nerves to stimulate their target tissue into action (or inaction).
- A few postganglionic neurons use acetylcholine, such as the nerves going to sweat glands to stimulate perspiration.
- The single neurotransmitter used in the parasympathetic nervous system is acetylcholine. Its effect on the body is the result of signal transduction from muscarinic acetylcholine receptors and their downstream targets.

NERVOUS SYSTEM DISEASES AND DISORDERS

Complexity Can Create Problems

Difficulties with the nervous system (either CNS or PNS) can lead to profound muscle dysfunction as well as cognitive, behavioral, and social disorders. While there are too many to explain in this section, some of the more familiar disorders are described.

Common nervous system disorders

Given the complexity of the nervous system, it's not surprising that occasionally communication breaks down and neurological disorders result:

- Epilepsy is a condition that occurs when abnormal electric signals from the brain create convulsions, affecting the way the body operates. Such seizures can be relatively minor or can be life-threatening.
- Stroke is caused by bleeding in the brain or a blocked blood vessel, and is often the result of high blood pressure, diabetes, or problems of the circulatory system.
- Peripheral neuropathy is caused by nerve damage, particularly peripheral nerves (hence the name), and can cause numbness and tingling in parts of the body.

Parkinson's Disease

Parkinson's disease results from lack of dopamine production or from a failure to detect this essential CNS neurotransmitter. Typically manifesting in adults over 50 years of age, symptoms begin with motor dysfunction including tremors, stiffness, and problems with walking. These early symptoms eventually lead to cognitive and social difficulties such as dementia as the disease progresses.

There is no cure for Parkinson's disease; much research has looked into disease prevention and management of the symptoms. Several studies have reported links between antioxidants (such as vitamin C) and the prevention of Parkinson's disease; however, this remains unresolved.

For patients who exhibit a decrease in dopamine production, medications that allow the neurotransmitter a longer active life are effective at reducing the symptoms. Typically, when dopamine is secreted toward synapses, an enzyme called monoamine oxidase (MAO) breaks down dopamine. MAO inhibitors reduce the efficacy of the enzyme, prolong the active life of dopamine, and mimic the presence of greater amounts of dopamine.

Alzheimer's Disease

Accounting for approximately 60–70 percent of dementia cases, Alzheimer's disease is a neurodegenerative disorder that primarily affects adults over the age of 65. Although the cause is poorly understood, heredity is thought to play a large role in determining who contracts the disease.

Several hypotheses have tried to explain the cellular/molecular events that lead to the disease; however, the leading hypothesis focuses on the role of amyloid plaque proteins (APPs) and their extraneous accumulation in the area surrounding neurons. It remains unclear if APPs cause Alzheimer's, since vaccines that remove the APP accumulations (plaques) do not reverse the dementia in these patients.

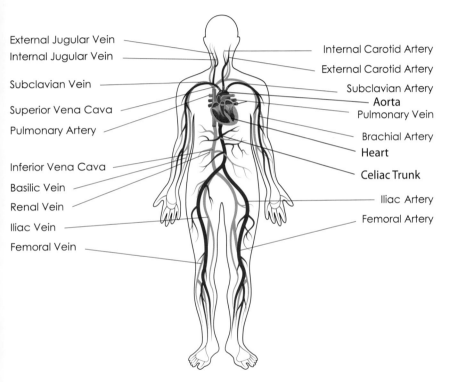

External Jugular Vein

Internal Jugular Vein

Subclavian Vein

Superior Vena Cava

Pulmonary Artery

Inferior Vena Cava

Basilic Vein

Renal Vein

Iliac Vein

Femoral Vein

Internal Carotid Artery

External Carotid Artery

Subclavian Artery

Aorta

Pulmonary Vein

Brachial Artery

Heart

Celiac Trunk

Iliac Artery

Femoral Artery

CARDIOVASCULAR SYSTEM

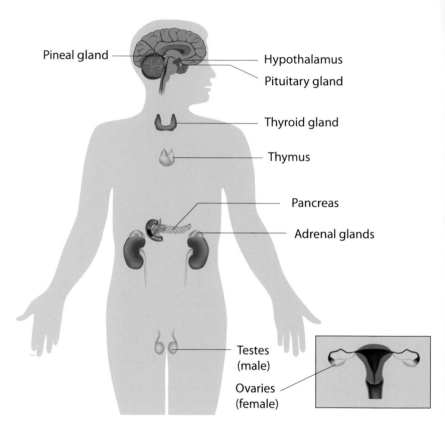

Pineal gland

Hypothalamus

Pituitary gland

Thyroid gland

Thymus

Pancreas

Adrenal glands

Testes
(male)

Ovaries
(female)

ENDOCRINE SYSTEM

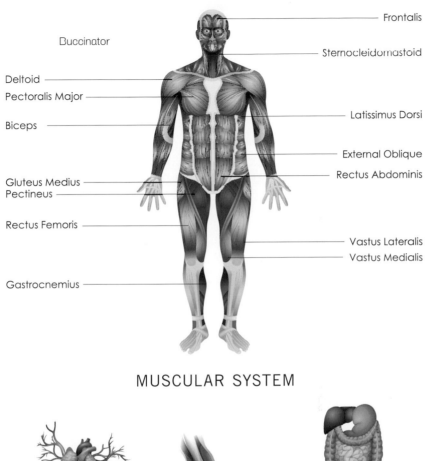

Frontalis

Buccinator

Sternocleidomastoid

Deltoid

Pectoralis Major

Latissimus Dorsi

Biceps

External Oblique

Rectus Abdominis

Gluteus Medius
Pectineus

Rectus Femoris

Vastus Lateralis

Vastus Medialis

Gastrocnemius

MUSCULAR SYSTEM

Cardiac muscle

Skeletal muscle

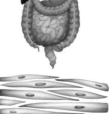

Smooth muscle

TYPES OF MUSCLES

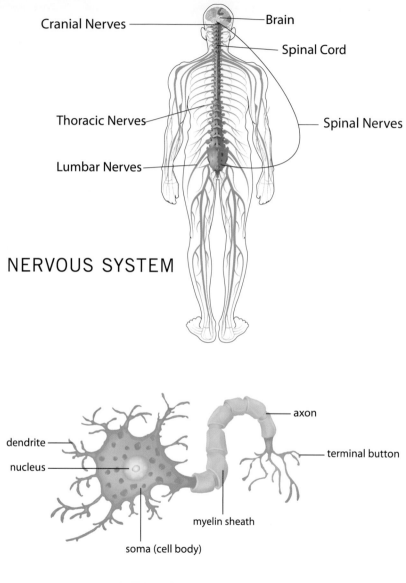

Cranial Nerves

Brain

Spinal Cord

Thoracic Nerves

Spinal Nerves

Lumbar Nerves

NERVOUS SYSTEM

axon

dendrite

terminal button

nucleus

myelin sheath

soma (cell body)

NERVOUS TISSUE

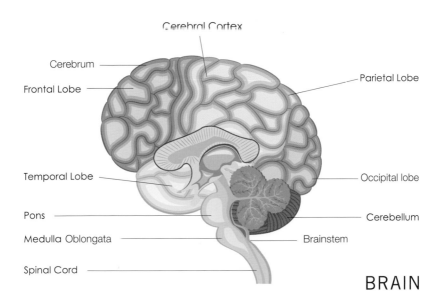

Cerebral Cortex

Cerebrum

Frontal Lobe

Parietal Lobe

Temporal Lobe

Occipital lobe

Pons

Cerebellum

Medulla Oblongata

Brainstem

Spinal Cord

BRAIN

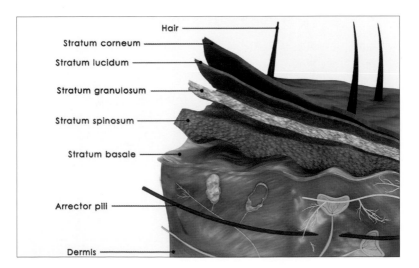

Hair

Stratum corneum

Stratum lucidum

Stratum granulosum

Stratum spinosum

Stratum basale

Arrector pili

Dermis

SKIN ANATOMY

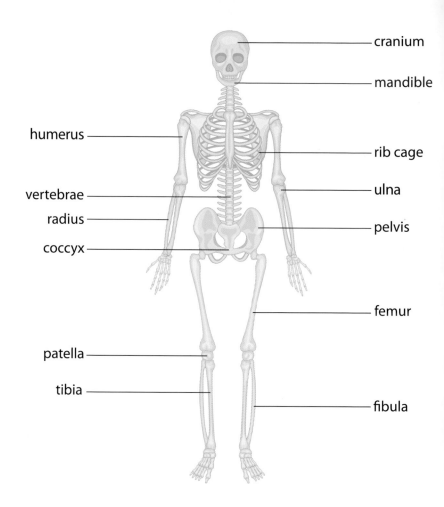

cranium

mandible

humerus

rib cage

ulna

vertebrae

radius

pelvis

coccyx

femur

patella

tibia

fibula

SKELETAL SYSTEM

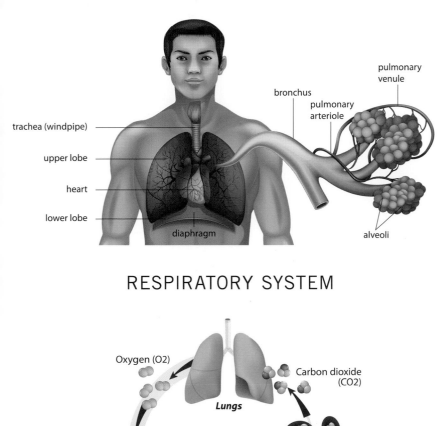

trachea (windpipe)

upper lobe

heart

lower lobe

diaphragm

bronchus

pulmonary
arteriole

pulmonary
venule

alveoli

RESPIRATORY SYSTEM

Oxygen (O2)

Carbon dioxide
(CO2)

Lungs

Red
blood cells

Organs

GAS EXCHANGE IN HUMANS

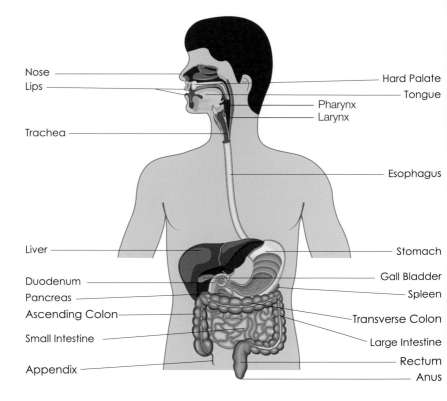

Nose

Lips

Trachea

Liver

Duodenum

Pancreas

Ascending Colon

Small Intestine

Appendix

Hard Palate

Tongue

Pharynx

Larynx

Esophagus

Stomach

Gall Bladder

Spleen

Transverse Colon

Large Intestine

Rectum

Anus

DIGESTIVE SYSTEM

Initial symptoms of Alzheimer's include short-term memory loss that, in time, leads to more severe cognitive disorders. Disorientation and problems with language can cause the individual to isolate himself or herself. Further advancement of the disease leads to loss of systemic bodily functions and death.

Accidental brain injury

Many injuries to the nervous system are caused by accidents and injuries. One of the most common of these is concussion, a brain injury that occurs with a sudden blow to the head (such as hitting your head against the steering wheel in a collision) and occasionally with a sudden blow to the body (such as being tackled during a football game). The brain moves abruptly (the Latin root of the word "concussion" literally means "to shake violently") inside the skull. This can cause loss of consciousness, swelling of the brain, or bleeding. Most concussions are minor and cause no lasting harm, but repeated or serious concussions can lead to brain damage.

Attention Deficit Hyperactivity Disorder

A neurodevelopmental disorder in children, attention deficit hyperactivity disorder (ADHD) or attention deficit disorder (ADD) affects a child's ability to focus attention for more than a few minutes on tasks with or without increased extraneous motor activity. These children are often identified once they begin socialized education (either preschool or primary school) because of their failure to stay on task and/or inability to sit quietly for any length of time. While some children diagnosed with ADHD or ADD simply may have not been socialized properly or taught acceptable behaviors at home, others do have a pathophysiological basis for these disorders and can benefit from counseling and medication. Stimulants are the most popular and effective means to treat children with ADHD or ADD, although this may not always be the best solution.

Autism

This disease begins to manifest itself in early childhood and can be seen in a delay in language development as well as social dysfunction. However, the most significant symptom is repetitive behaviors. These behaviors help in the distinction between autism and ADHD or ADD.

While many neuronal disorders have pathophysiological or chemical foundations, the underlying cause of autism isn't well understood. One possibility is the malformation of synapses throughout the CNS. This delays or incorrectly relays signals throughout the nervous system.

Many media articles have suggested a relationship between childhood vaccinations and autism. To date, the link between the two has been widely discredited within the medical and scientific community.

SENSORY SYSTEM RECEPTION AND PERCEPTION

Experiencing the World

The human body has an intricate system of sensors that provide information about the environment to the brain to elicit a response, which may be either conscious or subconscious, to provide better conditions for health and survival. The sensory system works 24/7 to keep your body functioning properly.

Reception and Perception

Obtaining information about the environment, whether interior or exterior, is a two-step process. The information must be obtained via specialized cells that can detect stimuli and it must be interpreted in context by the brain before a response can be made.

Receptors in your body provide a vast array of information to the brain such as internal and external temperature, blood pressure, light, sound, taste, and smell, as well as balance and body position. If any of these receptors fail, the information doesn't reach the brain and isn't detected. Therefore, no response, either reflex or conscious, is made.

The cause of "phantom limb"

While reception and perception generally require stimulus, the brain may have established patterns of dealing with sensory information that can continue even after loss of input. For example, amputees often experience sensation in the limb that is lost, something that is called phantom limb. Although the limb is gone, the brain still provides signals that make the person feel touch, pain, or temperature in that limb. This shows the power of the brain and its importance in the perception of a stimulus.

Information from receptors must be understood appropriately by the brain to be perceived. Tastes, smells, and images must be learned in the early years of life and assembled into a library of sensory information that can be sorted through when encountered again. If the brain never develops these early patterns because it was deprived of sensory input until later in life, the brain may never be able to accurately interpret the information and the individual may be unable to perceive certain stimuli. For example, individuals who have been blind from birth due to eye problems that are repaired later in life often have a difficult time with depth perception.

It takes both aspects of the sensory system working in unison for an individual to be able to adequately function and monitor the internal and external world in which we exist and survive.

Special Senses

The sensory receptors are categorized into groups based on how localized or widespread they occur in the body. Those that are localized in a single location are referred to as organs of special senses. This group includes the cells and tissues required for smell, taste, vision, hearing, and balance. In the past, touch was included among the "5" senses (instead of balance). However, much of the external surface of the body can detect touch as pressure. For that reason, touch was moved into the category of general senses.

General Senses

Unlike the special senses, receptors for the generalized senses are spread throughout the body. The detection of temperature, pressure, pain, and body position occur over much of the body surface and provide an abundance of sensory information to the brain.

Touch

You can find the receptors for touch between the upper layer of skin (epidermis) and the deeper layer (dermis). Some receptors are adept at detecting changes in pressure while others respond to vibrations or sustained pressure. Two of the more common receptors, Meissner's and Pacinian corpuscles, respond to slow and fast vibrations, respectively. Additionally, the Pacinian corpuscle detects deep pressure. These cells are specialized and encapsulated in such a way that the mechanical pressure leads to a signal transduction cascade and neuronal signaling. This is how the brain perceives touch.

How is pain detected?

Pain receptors (nociceptors) are free nerve endings that respond to various stimuli, which leads to the perception of pain. Extreme temperatures, excessive mechanical forces, and chemical damage are a few of the more common stimuli that are detected by these free endings as pain.

VISION

Now You See It, Now You Don't

This highly complex system is designed to detect light energy and transduce it into electrical information that is sent to the visual cortex of the brain for perception. Interestingly, light input in the right eye is perceived in the left part of the brain and vice versa. This crossing of the information and the slightly offset position of the eyes allows the brain to perceive different dimensions, including depth perception.

Anatomy of the Eye

The eyes are organs that capture and focus light energy on the back portion of the eyeball where special visual perception cells exist. Structures of the eye include:

- the retina, which contains the photoreceptor cells
- the sclera, which forms the outer covering of the eye
- the cornea (clear portion of the eye), situated just behind the sclera, which allows photons of light to pass into the eye
- the iris, found under the cornea, a spherical diaphragm that can open or close to regulate the amount of light that passes through
- the pupil, an opening in the iris, which is black because the back of the eye is dark
- the lens, which is suspended behind the iris and positioned immediately behind the pupil, and which focuses the light that enters the eye
- the ciliary muscle, which keeps the lens in place and can apply tension to the lens to stretch it or allow it to contract in order to

change the focal length of the light entering the eye (in order to shift between seeing objects up close and far away)
- the vitreous humor, which is the gelatinous material that fills the large chamber of the eyeball through which light passes after it enters the eye and is focused by the lens

Retina

The back of the eyeball has a layer of photoreceptor cells called the retina. Light passes through several layers of neurons and interconnecting cells before the photon impacts and excites a photoreceptor cell buried in the deeper parts of the retina. The only part of the retina deeper than the photoreceptors is a layer of pigmented epithelial cells that absorbs light and prevents its reflection to reduce extraneous light and improve fine detail.

How do some animals see at night?

Nocturnal animals have a reflective pigment in their epithelium that reflects light directly back onto the photoreceptor, thus stimulating it multiple times with a single photon and improving vision in low light.

Photoreceptors

These cells, found in the deep retina, contain a photo pigment that, when activated by a photon of light, change shape and lead to a signal transduction cascade, ultimately generating an electrical signal in a neuron that signals the visual cortex of the brain.

One type of photoreceptor is the rod cell. These cells are shaped much like a comb for your hair. Rows of folded membranes that resemble the teeth of the comb contain the photo pigment rhodopsin, which, when activated, leads to the change in the shape of the larger molecule where it is attached (opsin). An essential component of this photo pigment complex is a molecule called retinal, which is related to vitamin A. Rods are the more sensitive of the two types of photoreceptors and are responsible for vision in low light, which many have referred to as black-and-white vision.

Why do people say eating carrots improves your eyesight?

Carrots are a rich source of vitamin A, which is part of the photo pigment complex.

Cones are the second photoreceptor and are responsible for color vision in the retina. These cells also have folds of photo pigment, but are shorter than the rods and have a tapering or cone shape, hence the name. Red, green, and blue cones detect different wavelengths of light.

Which cones are stimulated and to what degree determines the color of light that is signaled and perceived. This is not unlike your television screen or computer monitor where these three same colors are used to blend into any visual color in the spectrum. While cones are much less abundant than rods in the retina in general, in the fovea of the retina (the central part of the retina responsible for sharp central vision), cones greatly outnumber rods.

HEARING, BALANCE, SMELL, AND TASTE

The Fab Four

The two special senses of hearing and balance use mechanical energy to perceive sensation, while the two special senses of smell and taste detect chemical signals.

Hearing

Sound waves (mechanical energy) move through the air and are funneled into the auditory tube via the external ear (pinna). This canal and the eardrum (tympanic membrane) at the deep end make up the outer ear. As waves impact the eardrum, the membrane vibrates, and energy that is transferred through a series of small bones in the middle ear, the chamber on the opposite side of the eardrum.

The small bones in the middle ear are the malleus, incus, and stapes (also called the hammer, anvil, and stirrup, respectively). They transfer the mechanical energy of the sound waves from the eardrum to the cochlea, the organ of the inner ear that houses the auditory receptors. The stapes is attached to an oval window on the body of the cochlea. Waves are transferred into the fluid-filled tunnel of the coiled cochlea.

The cochlea resembles a snail shell. The long fluid-filled tube, through which the sound waves pass, is packed inside. The tube is divided into an upper and lower chamber by the sensory organ of Corti. The very end of the tube remains open and allows the sound waves in the upper chamber to run the full length of the tube. In doing so, the waves coursing through the fluid compress a membrane in the organ of Corti, which then compresses down upon receptor cilia to cause an electrical signal.

These receptor cilia, like the organ of Corti, run the full length of the tube. The area in the cochlea where the receptors are stimulated corresponds to the wavelength of the sound and distinguish low from high tones. Think of depressing the key on a piano. Depending on where on the keyboard you push, a different sound will emerge.

Frequency range in humans

Humans can't perceive all sound wavelengths. Individuals with normal hearing can detect sound in the frequency range of 20 Hz (hertz) to 20,000 Hz. (A hertz is 1 cycle per second, with a cycle being how often an alternating current changes direction, or frequency.) Many animals can hear sounds that humans cannot hear, such as ultrasound (higher than 20,000 Hz) and infrasonic (lower than 20 Hz). Dogs can detect ultrasound, which is why they can hear the mail truck coming long before you can.

Balance

Balance, which was added in recent years to the list of special senses, also has receptors located in the inner ear and is associated with the cochlea. While sound is detected in the body of the cochlea, balance is monitored in three semicircular canals attached to the body of the cochlea. Oriented in three planes, which detect movement in three dimensions, these canals respond to acceleration and deceleration and use the law of inertia to function.

The law of inertia

Also called Newton's first law of motion, the law of inertia simply says that an object at rest tends to stay at rest and an object in motion tends to stay in motion unless acted on by an opposing force.

Fluid called endolymph fills each semicircular canal, and much like any object, moves relative to the body and follows the law of inertia. Receptor cells project cilia upward and into this fluid. At rest, the cilia stand straight and no signal is sent to the body. However, when the body starts to move, the hair cells (and cilia) accelerate at the same rate as the body. But the endolymph remains stationary for a brief moment—just long enough to bend the cilia of the receptors and transduce an electrical signal that motion has been initiated. Likewise, when the body stops, the fluid remains in motion and bends the cilia in the other direction. The brain perceives this as halting.

Smell

The sense of smell, or olfaction, begins in the lining of the nasal cavity. Secretory cells coat the surface of the nasal epithelium with a watery, protein-rich fluid that functions to trap chemicals as they are brought into the nasal cavity during an inhalation.

Once they are trapped in the watery surface zone, these chemicals are detected by the olfactory receptor cells, which resemble neurons more than epithelium. On their surface and projecting into the watery, chemical-filled fluid are dendrites, often called cilia. The cilia increase the surface area where proteins can bind to the chemicals and begin a signal transduction pathway that leads to an electrical signal that is sent to the brain. At the base of these cells, axons transmit the electrical signals on to mitral cells located in what is called the olfactory bulb, which is connected to the olfactory centers of the brain via the olfactory tract.

The olfactory tract is often incorrectly called the olfactory nerve. However, the axons of the olfactory receptors collectively form the olfactory nerve proper (filia olfactoria).

Taste

Taste (gustation) is the second of the chemically mediated specialized senses and is initiated on the dorsal surface of the tongue via barrel-shaped structures called taste buds. Located on the lateral portions of specialized papillae (projections), the taste buds are a collection of supportive and receptor cells used for taste. Hairlike microvilli from the receptor cells project into this opening to collect the chemicals dissolved in the saliva on the surface of the tongue. These receptor cells form synapses with sensory neurons at the base of the taste bud and transmit information away from the tongue and on to the taste center of the brain.

Taste buds are capable of detecting several chemical stimuli, but are thought to specialize in one or two. They are distributed across the surface of the tongue and are restricted only by the locations of the papillae on which they are located. Receptors meditate five basic tastes that require distinct signal transduction mechanisms to convey the stimulus and transduce it into electrical energy. Sweet, salty, sour, and bitter are four of the basic tastes that have been identified for decades. While most are familiar with these, the most newly identified of the basic tastes is umami, which is a savory flavor and common in Asian cuisines.

SENSORY SYSTEM DISEASES AND DISORDERS

When the Signal Doesn't Get Through

The complex and intricate systems that detect and respond to stimuli can occasionally break down or suffer problems. Injuries, infections, and cancers can affect the senses. Other sensory system diseases and disorders include:

- sensory processing disorder, where the brain receives stimuli but doesn't organize it in a coherent way
- age-related macular degeneration, where blood supply to the retina is disrupted, causing loss of vision
- nasal polyps, benign (noncancerous) growth in the nose, often caused by allergies or infections

Color Blindness

This condition results from an inadequacy, malfunction, or malformation of one or more types of cones in the retina. The most common type of color blindness is red-green color blindness, when red and green color discrimination is decreased. Affecting about 1 in 12 males, this is a hereditary condition and is termed X-linked, meaning the gene that is the problem is on the X chromosome. This is a recessive mutation. The presence of a dominant gene can overcome the condition and the individual will only be a carrier of the color-blind gene.

Anosmia

Anosmia means a lack of the sense of smell. Smell and taste are intimately linked together to expand the repertoire of each sense. A person lacking the sense of smell has a much less discriminating palate when compared to individuals with a functional olfactory system. Often, temporary anosmia may occur during a sinus infection, and that is usually why the taste of food is dulled during such a sickness.

Vertigo

Vertigo is a result of complications with the functioning of the inner ear, specifically the semicircular canals. The lack of information, or incorrect information, pertaining to balance can result in extreme dizziness, nausea, and vomiting, and increases the likelihood of falling. It is particularly problematic in older individuals.

Motion Sickness

With some of the same symptoms as vertigo, motion sickness (also called car sickness or air sickness) results from crossed signals in the visual centers of the brain. There is confusion between what the brain thinks is happening versus what the balance centers of the brain think is happening.

For example, many people get seasick on a ship when they are inside of a portion of the ship that has no windows. Their visual cortex sees the floor, walls, and ceiling, and relative to the body there is no movement. However, because the boat is moving up and down with the waves, the body is in fact moving and this fact is being detected by the semicircular canals of the inner ear.

CARDIOVASCULAR SYSTEM AND HEART STRUCTURE

Have a Heart

The cardiovascular system transports materials throughout the body that cells and tissue could not survive without. Conveying oxygen to the tissues and relaying carbon dioxide back to the lungs is an essential function of this system. However, other critical materials are also conveyed in the blood, such as hormones and immune system cells. The cardiovascular system is much more than merely a pump and plumbing to transport gases.

Heart Structure

The driving force of the cardiovascular system is the heart. This muscular organ begins pumping even before the heart is fully formed in the embryo and consistently and spontaneously beats for a lifetime. Divided into four chambers and separated into right and left portions, the heart functions as a pump for two circulations: the systemic circulation for the body and the pulmonary circulation for the lungs. Using both of these removes carbon dioxide from the body while supplying fresh oxygen to the tissues.

Chambers

The two superior (upper) chambers of the heart are the atria, which actually begin as a single chamber during embryonic development and later form a partition (interatrial septum) to divide into right and left chambers. These thin-walled chambers are the primary holding areas for blood returning from the body (right atrium) and from the lungs (left atrium). Movement of the blood from the atria downward into the respective ventricles (lower chambers) is accomplished primarily from low pressure pulling the blood downward.

Like the atria, the ventricles begin as a single chamber that is separated into right and left chambers by a thick muscular partition (interventricular septum).

How the heart's pump system works

Oxygen-depleted blood in the right atrium is pulled into the right ventricle as its walls expand during the relaxation phase (diastole) of a cardiac cycle. Likewise, oxygen-rich blood in the left atrium is pulled into the left ventricle. The left ventricle has the thickest of the muscular walls because the blood in this side must be pumped throughout the body and back to the heart, while blood in the right ventricle must only have sufficient pressure to make it to the lungs and back to the heart.

Cardiac Valves

The cardiac valves regulate the passage of blood from one chamber to the next as well as ensure that the blood only flows in one direction. Atrioventricular valves regulate the movement of blood between the atria and ventricles. The structure of these valves is similar in both the right and left atrioventricular canals. Formed from thin connective tissue sheets (cusps), the tricuspid valve (on the right) and the bicuspid valve (on the left) are pushed against the wall of the ventricles as the blood is pulled into the ventricles from the atria. When the ventricles contract (systole), blood is compressed and forced upward in the ventricles. The most open path out of the ventricle is back up into the atrium.

As the blood moves up, it is forced under the cusps, which are inflated, and moved upward toward the atrium. If the cusps were the only structural portion of the valves, they would simply be pushed aside into the atrium as the blood rushes upward. This does not happen because attached to each cusp are cords (chordae tendineae) made from tough connective tissue. These cords are anchored into large bundles of cardiac muscle (papillary muscle) present in the ventricle that contract as the ventricle as a whole contracts.

Thus, as the blood moves upward because of the ventricular contraction, the papillary muscles pull downward on the cords with the same force as the upward moving blood. Because these forces are equal, they allow the cusps to move upward until they are aligned at the atrioventricular canal. This closes off the canal, preventing the blood from moving backward (regurgitation). The closing of these valves produces the first heart sound, or the "lub" that is heard for the heartbeat.

Another set of valves, the semilunar valves, prevent blood from flowing back into the heart from major vessels. These valves are present in the great vessels just outside of the heart. Blood from the right ventricle is forced into the pulmonary trunk (a major blood vessel that divides into the two pulmonary arteries) and is destined for the lungs. Blood from the left ventricle is moved into the aortic trunk (the root of the aorta that attaches to the heart) and is pushed to the rest of the body. Located within each of these vessels are three cusps that resemble pockets. These cusps inflate when the blood falls downward in an attempt to return to the ventricles, effectively closing off the passageway and keeping the blood in the vessel. The semilunar valves allow blood to exit the ventricles, but not return.

HEART REGULATION

With Every Beat of Your Heart

A unique feature of the heart is that it beats spontaneously and rhythmically on its own. Regardless of any nervous system signals, the heart starts beating before you are born and consistently and continually beats until your death.

Pacemaker and Conduction

Setting the beat for the heart is the task of the pacemaker tissue in the right atrium. Much like a metronome continues to click once started, the pacemaker, once generated, beats for a lifetime. Beginning as tissue in the embryonic sinus venosus, a primitive and embryonic chamber receiving blood from the body, it becomes incorporated into the right atrial wall as development proceeds. The pacemaker is named the sinoatrial node (SA node) to represent its embryonic origins and its final adult location.

Cells of the sinoatrial node are modified cardiac muscle connected to the muscle cells of the atrium via the gap junctions of the intercalated disks. (Gap junctions allow direct cell-to-cell contact; intercalated disks are features of cardiac muscle that allow for contraction.) When the cells of the SA node spontaneously generate an action potential, it spreads to all the cells of both atria through these junctions.

The muscles of the atria and the muscles of the ventricles are separated by a connective tissue ring called the annulus fibrosus that forms the foundational anchor for the valves and the septa of the heart, which prevent the SA node signal from spontaneously spreading to the ventricles. Additional conductive cells pick up and relay the electrical signal through the annulus downward to the ventricles.

Anatomy of a Word
septa

Septa is the plural form of septum, and a septum is a wall. In the heart, septa separate the four chambers of the heart. The septum that divides the left and right atria is the atrial septum or the interatrial septum. The interventricular septum (or ventricle septum) divides the ventricles.

In the right atrioventricular region, not far from the tricuspid valve, is another area of modified cardiac muscle called the atrioventricular node (AV node). The AV node cells are connected to the atrial muscle via gap junctions (just as the SA node is connected to the atrium), so they are stimulated by the spreading electrical signal from the SA node. After detecting the signal, the AV node pauses before relaying the signal to the ventricles. This allows the atria to contract just before the ventricles contract, and therefore enables the ventricles to fully fill with blood before the next contraction.

After pausing, the AV node cells relay the signal through modified muscle cells arranged into a bundle of fibers that run through the annulus fibrosus and down toward the apex of the ventricles. This AV bundle (bundle of His) transfers the electrical signal to the base of the heart, where the fibers spread throughout the ventricle and regenerate the electrical signal in the ventricular muscle leading to the contraction.

EKG

The health and state of cardiac function can be assessed indirectly and noninvasively by detecting the electrical changes that occur in the heart. Electrodes positioned on the chest can receive the electrical potentials and display them as a series of peaks or waves that correspond to the electrical activity in the different chambers of the heart. This display is termed an electrocardiogram (EKG).

Where does the acronym EKG come from?

EKG is used to designate an electrocardiogram because, in German, the heart is referred to as Kardia. Also, this convention helps prevent confusing EKG with other diagnostics test such as an echocardiogram (ECG) or an electroencephalogram (EEG).

The initial small peak seen at the start of a cardiac cycle is the P wave, and represents the depolarization of the atrial muscle that precedes the contraction of the atria. The next wave is the largest and sharpest. This is the QRS complex, named for the bottom starting point (Q), the top of the spike (R), and the bottom point (S) of the wave that occurs during ventricular depolarization (the depolarization that precedes the contraction of the ventricle). Following the QRS complex is an intermediate-sized wave, the T wave, and it occurs during repolarization of both the ventricles and the atria. Any changes in the size of the waves or the timing between peaks has diagnostic value for the cardiologist.

Heart Rate

Although the heart is capable of beating on its own, it does not possess the ability to know when to increase or decrease its rate based on activity level. This is the role of the autonomic nervous system. During heavy exercise, your heart must increase its rate to provide the needed materials including oxygen to the hard-working muscles. Neurons produce and secrete norepinephrine and the adrenal gland produces epinephrine (adrenaline). These molecules cause the cells of the pacemaker to increase the rate of firing and lead to an increase in overall heart rate. Conversely, during periods of inactivity, such as sleeping, other neurons secrete acetylcholine, which causes the pacemaker to slow down and decreases the rate.

Strength of Contraction

In addition to beating faster, the heart may contract harder to eject more blood per beat during periods of increased demand. This strength of contraction doesn't depend on outside signals but is built into the cardiac muscle itself. During periods of normal activity, the actin and myosin filaments overlap in such a way that not every actin can form a cross bridge with myosin. Only as the heart increasingly fills with blood and the fibers are stretched farther apart can the actin fully engage the myosin and provide the most intense contraction. In this way, the design of the muscle fibers provide a built-in reserve to be used when the heart is filled with more blood, which is during times of increased demand and results in more blood being pumped out of the heart.

BLOOD VESSELS

Your Body's Highway System

The vascular system consists of blood vessels that permeate the human body and convey all essential materials, both cellular and molecular. Some vessels are subjected to high blood pressure (arteries), which their structure must resist, while others (veins) must assist the extremely low-pressure blood in returning to the heart. Other vessels are only one red blood cell in diameter, which allows the direct exchange of materials between the blood and the body's tissues.

How extensive is the vascular system in the human body?

Connected end to end, all of the blood vessels in the body would extend approximately 60,000 miles. For reference, the circumference of the earth at the equator is just under 25,000 miles.

Arteries

The vessels that transport blood away from the heart are called arteries. Not all of the blood carried in arteries is oxygen-rich; the pulmonary arteries transport oxygen-depleted blood from the right ventricle to the lungs. Arteries, like all types of vessels, have a lining layer of epithelium called the endothelial layer. These cells, because of their overall net negative surface charge and molecular composition, provide a low-friction surface for blood cells and platelets to flow across.

The three layers of an artery include:

- The endothelium (the name of the epithelial lining, or endothelial layer, of vessels) and the underlying connective tissue is grouped together into a layer termed the tunica intima. This is the innermost of three layers present in most arteries and veins.
- The middle layer, which is most prominent in arteries, is the tunica media and consists of varying numbers of smooth muscle cells and sheets of elastic fibers (elastic laminae). Together the elastic component (which rebounds following high pressure) and the smooth muscle cells (which control the narrowing or expansion of the vessel diameter) assist in maintaining both blood pressure and the flow of blood through different vessels.
- The outermost layer of the vessels is the tunica adventitia and is composed of the connective tissue layer surrounding the vessel.

The largest-diameter arteries, such as the aorta, are defined as elastic arteries because of the large number of elastic layers present in the tunica media. Pressure is highest when the blood immediately leaves the heart, and these vessels require the elastic fibers to resist and rebound to the high-pressure stretching. Most of the intermediate-sized vessels have layers of smooth muscle in the tunica media that may reach up to 40–50 layers thick.

As blood flows farther away from the heart, the layers of the arterial wall become fewer and the diameter of the vessel decreases until all that remains of the tunica media is one or two smooth muscle cells along with the tunica media. These are called arterioles (small branches of an artery) and immediately precede capillaries, tiny vessels that make up the body's microcirculation. Often the smooth muscle of these vessels acts as a pressure regulator to prevent the blood pressure from being too high before entering the capillary network.

Capillaries

The capillaries are the small vessels that allow the direct exchange of materials between vessel and tissue. They are composed of a single endothelial layer with little, if any, connective tissue.

Most of the capillaries in the body are categorized as continuous capillaries. The endothelial cells of these capillaries form tight complexes between cells and only material passed through the cell is transported. There is no place in the body where this is more evident than in the brain, where the continuous capillaries are a part of the blood-brain barrier. In other areas, material needs to be transported more rapidly and the specificity (e.g., type of molecule being transported) isn't critical. The endothelial cells in areas such as in the kidney have small pores in the endothelial cells that allow larger material to be transported more quickly. These openings, called fenestrae, may be open or covered with a thin membrane to further specify (that is, define the type of material) what can pass. In either case, these capillaries are called fenestrated capillaries. The final type of capillary is somewhat like Swiss cheese in that there are almost more holes than cells in these vessels. In organs such as the liver, in which cells are in contact with much of the plasma component of the blood, only the cells are restricted from access to open areas called sinuses or sinusoids. All other material can pass through these large holes. Thus, these are termed sinusoidal capillaries.

Veins

By the time the blood passes through the capillaries, the blood pressure has been drastically decreased. In the aorta, the blood pressure was approximately 100 mmHg. When the blood finally returns to the heart, the pressure falls to almost 0 mmHg. The exchange of materials in the capillaries and their small diameter allow this low-pressure blood to enter and return back to the heart.

What does mmHg mean?

In medicine, mmHg is used as a measurement of force per unit. It is based on an old method of comparing the amount of pressure that mercury (a heavy metal) exerts to other types of pressure (such as blood pressure): 1 mmHg is the amount of pressure generated by a column of mercury 1 millimeter high. It is the equivalent of 1/760 atmosphere, an atmosphere being a unit of measure equaling about 14.7 pounds per square inch.

Vessels called venules are primarily a single endothelial layer. However, their diameter is larger than that of capillaries. They usually are found adjacent to an arteriole. From the venules, the blood moves into increasingly larger-diameter veins. This increase in diameter causes the blood to pool and also decreases its pressure.

Unlike arteries, the thickest layer of veins is the adventitia, or the outer connective tissue layer. Little smooth muscle is present in the tunica media of veins. Another identifying characteristic of veins is their diameter, which is much greater than the thickness of the vessel wall (the opposite is true for arteries). Since the blood flow in the veins lacks sufficient pressure to return blood to the heart, veins possess unidirectional valves. With each contraction of the heart, blood is moved upward through a set of valves, which then close as the heart relaxes, keeping the blood from falling backward due to gravity.

CIRCULATION

How Your Blood Gets Around

Moving the blood from heart to lungs and from heart to body and returning it back to the heart is a monumental task that requires miles of vessels throughout the body. In this section is a list of some of the major arteries and veins that should be familiar to anyone looking for a better understanding of the vascular system of the human body.

Major Arteries

As blood leaves the ventricles of the heart, it will pass into one of the so-called "great vessels." From the right ventricle, the blood moves into a single large pulmonary trunk (the aorta) that soon divides into right and left pulmonary arteries on the way to the lungs. From the left ventricle the blood enters into the aorta that arches up over the superior portion of the heart where large vessels leave to service the superior part of the body and the arms.

Anatomy of a Word

brachiocephalic

Brachio means "arm" and *cephalic* refers to the head; hence the name of the artery that first branches off the aorta, the brachiocephalic, indicates that it goes to the head and arm.

The first branch from the aortic arch is known as the brachiocephalic artery, which sends blood to the right side of the body including head and neck. It branches into two arteries, the right common carotid artery and the right subclavian artery, which extends through the body wall and into the arm as the brachial artery.

Just beyond the brachiocephalic artery, two additional arteries extend directly from the aortic arch, the left common carotid and the left subclavian artery. The aorta continues to arch 180° and descends into the body as the systemic or descending aorta, from which other arteries branch to the remainder. The largest branch off the aorta is the celiac artery, which supplies most organs of the upper digestive tract. From the celiac, many arteries branch out:

- The artery that goes to the liver is the hepatic artery.
- The artery that goes to the stomach is the gastric artery.
- The artery that goes to the spleen is the splenic artery.
- The artery that goes to the right and left kidneys are the renal arteries.

Moving farther down, the superior and inferior mesenteric arteries sprout from the aorta and supply portions of the small and large intestines, respectively.

At the most inferior end, the aorta divides into the right and left common iliac arteries that descend into each leg, becoming the external iliac arteries in the pelvic region. The vessels continue down each leg as the femoral arteries, which branch to the muscles of the leg and the foot.

Major Veins

Many of the major veins in the body mirror those of their counterparts of the arterial circulation. Blood from the feet and legs are drained from tissue via the femoral vein into the common iliac. This moves blood into the inferior vena cava (also called the posterior vena cava), the large vein that transports deoxygenated blood back to the right atrium. The inferior vena cava supplies the major venous drainage for the abdominal and inferior part of the body.

While many of the abdominal veins are named as their arterial counterparts (superior and inferior mesenteric, gastric, splenic), the majority of the veins from organs of the digestive tract drain into a large vein called the hepatic portal vein, which transports the blood to the liver.

In the liver, blood passes through the sinusoidal capillaries to the liver cells where it is metabolized. Here materials can be processed, excreted, or stored. From these sinusoids, blood is moved out of the liver by the hepatic vein into the inferior vena cava.

Blood from the head and neck is drained by large veins called the internal and external jugular veins, which empty their blood into the arching right or left brachiocephalic vein. These join together in the midline of the body to form the superior vena cava, which returns blood to the right atrium. From the arms, blood moves upward into the brachial vein, which becomes the subclavian vein and joins onto the brachiocephalic to return to the heart.

What is blood pressure?

Hydrostatic pressure (blood pressure) is the pressure the blood exerts on the vessel walls due to the contraction of the ventricles of the heart. With each ventricular contraction (systole), high-pressure blood is ejected into vessels of the body. For the average person, the pressure in the brachial artery during this contraction will be approximately 120 mmHg, and is termed the systolic pressure. During a ventricular relaxation (diastole), when the ventricle is expanding and creating a lower-pressure area to pull blood from the atria, the pressure will fall to near 0 mmHg inside the ventricles. However, because of the similar valves in the great vessels and the elastic nature of these arteries, the blood pressure will not fall below 80 mmHg on average and represents the person's diastolic pressure.

Cardiovascular System Diseases and Disorders

Any problem that causes the pumping action of the heart to be diminished or restricts blood flow into specific areas of the body can lead to reduced life expectancy or imminent danger of death.

Ischemia

Because of diet or heredity, many people have high cholesterol levels in their blood streams. Over the long term, if left untreated, this material can accumulate in the walls of blood vessels and obstruct the flow of blood, a condition called ischemia. If the narrowed vessel is in the heart, lungs, or brain, the danger may be grave. Some lipoproteins in the blood help to remove these cholesterol accumulations in the blood vessel, while others actually make the situation worse. Often referred to as "good" or "bad" cholesterol, these lipoproteins can be tested for and adjustments made to lifestyle and diet to help improve ischemia.

Myocardial Infarct

If an ischemic event occurs in the vessels supplying the heart itself, the cardiac muscle, which is not designed to function under hypoxic (low-oxygen) conditions, can be destroyed, causing a heart attack. A heart attack is actually muscular death and degeneration of the heart muscle. It does not regenerate, but is replaced with connective tissue.

As one can imagine, if enough muscle dies, the heart either functions poorly or doesn't function at all. One vessel that is critical to heart function and must remain open is the left anterior descending (LAD) coronary artery. It supplies two-thirds of the left ventricular muscle and if it becomes blocked, a person will suffer a massive heart attack, from which survival is unlikely.

RED BLOOD CELLS

The Taxis of the Human Body

While the cardiovascular system may be thought of as the interstate highway system of the body, the blood and its many components are most definitely the vehicles, carriers, transporters, and workers that keep the body supplied with essential materials. Additionally, some of these components act as garbage trucks, removing waste and keeping tissues healthy.

Cellular Function

The oxygen-carrying red blood cells (RBCs, also called erythrocytes) are packed with the oxygen-binding molecule hemoglobin, which give the cells their red color. They shuttle oxygen from the lungs to the tissues of the body where it can be used. RBCs are essential for the processing of CO_2 and its transportation in the plasma as bicarbonate.

Cell Formation

Erythropoiesis is the process of forming RBCs from precursor cells. In an adult, RBCs are produced in the marrow of the long bones. However, in some instances, the liver and even the spleen may function in erythropoiesis.

The number of RBCs in the blood remains fairly constant as the newly formed RBCs equal the number of old RBCs removed from circulation daily. This balance is under close hormonal control to ensure that the numbers do not increase or decrease beyond a functional range.

When the body detects decreases in the amount of available and transported oxygen (hypoxia), the hormone erythropoietin (EPO) is released from the kidneys and leads to increased synthesis of RBCs in the bone marrow. These additional cells aid in the transportation of sufficient oxygen to the tissues. Once the number of RBCs is balanced with available oxygen and oxygen use in the body, EPO levels decrease and RBC formation returns to normal levels.

Cell Structure

As RBCs develop in the bone marrow, the cells become progressively smaller and redder in color as the cytoplasm fills with hemoglobin. Toward the end of their development, round RBC precursor cells called orthochromatophilic erythroblasts abandon their nucleus, which allows the cytoskeleton of the RBC (now called a reticulocyte) to adopt its typical and final bi-concave shape.

This shape maximizes the number of hemoglobin molecules that can bind to oxygen. If the cell remained round, the molecules at the very center would never encounter oxygen because they would be too far away from the plasma membrane.

Hemoglobin

Hemoglobin is made of 4 protein molecules joined together. Adult hemoglobin is composed of 2 alpha and 2 beta chains, each of which has an amino acid structure called a heme group capable of binding to a molecule of inorganic iron. The iron molecule, in turn, binds to oxygen in a reversible manner (that is, the oxygen can attach and release). Therefore, each hemoglobin molecule binds to 4 molecules of oxygen. When oxygen is bound to hemoglobin (oxyhemoglobin), RBCs turn a red color. Conversely, when oxygen is unloaded from hemoglobin (deoxyhemoglobin), the cells turn bluish in color.

Destruction

As RBCs age and approach their 120-day life expectancy, their plasma membrane becomes more rigid and the cell as a whole is less flexible. Recall that a capillary is only about the diameter of a single RBC (8 micrometers). Therefore, RBCs must be flexible enough to squeeze through capillaries in single-file fashion. Otherwise they could block the capillaries. As the cells circulate, they pass through the spleen where they must squeeze through barrel-shaped sinuses. Healthy RBCs traverse the spaces of the sinusoids effectively and in the process have debris cleaned from their surfaces. However, older, rigid RBCs are shredded as they are forced through the narrow space, and are thus destroyed and removed from circulation.

The spleen also contains an abundance of resident macrophages, the vacuum cleaners of the body. These phagocytic cells remove the cellular debris and waste material.

Blood Groups

Marker proteins and carbohydrates on the surface of red blood cells are formed into groups, which allow for identification and matching of blood cells from donor to recipient in clinical and emergency cases.

ABO Group

The most familiar of the blood groups, and the one which most refer to when asked their "blood type," is the ABO blood group.

All RBCs have the same foundational protein on their cell surface. This base molecule is called the H antigen (an antigen is a molecule that can be recognized by the body's immune system). If no additional carbohydrate modifications to this H antigen are present, these cells form O-type blood. A person with A-type blood will have an additional n-acetylgalactosamine on the H antigen, while a person with B-type blood will have an additional galactose carbohydrate on the base H antigen. Since ABO genetics is a form of codominant inheritance, a person with AB blood type would have some H antigens with the A-type carbohydrate, while other H antigens would have the B-type carbohydrate. There would never be both carbohydrate antigens on the same H antigen.

How can antibodies develop against an antigen an immune system has never encountered?

Many scientists have attempted to explain this problem by hypothesizing that either environmental agents similar to the blood groups or viruses that deliver these antigens infect every human in the early months and years of childhood and stimulate the production of these antibodies.

Rh Factor

The rhesus (Rh) blood group is so named because of the monkeys in which it was first identified. Rather than being a single antigen, several different genes can be expressed on the red blood cell surface and result in a person being Rh positive (the most common Rh antigen is RhD). In fact, the vast majority of the human population is Rh positive. Only when no Rh antigen is on the RBC surface is the blood considered Rh negative. Antibodies against Rh factor will only be produced if and when the blood of an Rh-negative person comes into contact with Rh-positive blood. In normal life, this would be a rare event for most people. However, for Rh-negative females who may become pregnant with babies carrying the Rh factor, it could present complications.

WHITE BLOOD CELLS

The Infection Fighters

Although they make up less than 1 percent of blood volume, white blood cells (also called leukocytes) function as essential players in the immune system and aid in clearing cellular and pathogenic debris from the body.

Leukopoiesis

Stem cells for WBCs are present in the marrow of bones and through a series of developmental stages mature and become specified for their particular function. However, many WBCs, especially lymphocytes (cells that determine the specific nature of a foreign substance), don't fully mature in the bone marrow, but instead develop in lymphoid tissue, such as the thymus, spleen, and lymph nodes.

During this maturation time, the WBCs distinguish themselves based on the presence (or absence) of specific granules. Those cells with granules (which contain enzymes that destroy microorganisms) are collectively called granulocytes, and those without are agranulocytes.

Granulocytes

The cells in this group are all phagocytic cells, meaning these cells will engulf pathogens and debris with their plasma membrane. Once the cellular membrane captures this material, vesicles fuse with the captured material and its contents. Some of these vesicles, called lysosomes, contain hydrolytic enzymes that break down the pathogen or debris contained within the membrane compartment. These cells have the ability to follow chemical trails to the site of pathogenic attack via a process called chemotaxis. These cells can leave the circulation to mount an immune attack in the tissues of the body.

Anatomy of a Word

diapedesis

Diapedesis is the process of WBCs squeezing out of the endothelial cells of capillaries and gaining access to the tissue compartments.

Neutrophils

Neutrophils, a type of granulocyte, are the most abundant of all WBCs and make up about 60-70 percent of the WBC population in healthy individuals. These cells live, on average, about 12 hours and are always the first on the scene of an infection. They are also the first to die as they engulf a pathogen and destroy it along with themselves. Another distinguishing characteristic of the neutrophils is their multi-lobed nucleus. As the neutrophil ages, more and more lobes will appear on the nucleus, with a maximum reaching approximately five lobes. With this difference in appearance, these cells are also often referred to as polymorphonuclear (PMN) leukocytes.

Eosinophils

Eosinophils represent 2.5 percent of all WBCs. These phagocytic cells destroy parasitic organisms and function during allergic reactions. Although they are the same size overall as neutrophils, these have larger granules. Often, the granules are so large that they obscure observation of the bilobed nucleus of these cells.

Basophils

Basophils are the final member of the granulocytes and make up less than 1 percent of WBCs in the circulating blood. These cells have the largest granules, which prevent the direct observation of the nucleus. Basophils are also the only member of the granulocytes that adopt a different name when they leave circulation. In the tissues of the body, they are called mast cells. Their granules contain many inflammatory mediators, including heparin (which inhibits blood clotting) and histamine (which increases vascular permeability).

Agranulocytes

The remaining WBCs that lack granules make up both the largest and the smallest of the WBCs. Monocytes are the largest of the WBCs, about three times the size of an RBC, and make up about 8–10 percent of the WBC population. These are the vacuum cleaners of the body. Highly phagocytic, they move in and out of the circulation, capture and destroy pathogens, and return to areas populated with other WBCs (lymph nodes, thymus) and present antigens to these immune cells to initiate an immune response. Like the basophils, monocytes also change their name when they leave the circulation.

The smallest and the second most abundant of the WBCs are the lymphocytes. Lymphocytes are organized into two groups. B lymphocytes, which start their maturation in the bone marrow, are the cells responsible for the humoral (antibody-mediated) immunity of the body. T lymphocytes, which mature in the thymus, function in the cell-mediated immune system. Both types of lymphocytes are about the same size as RBCs and are identified as having very little cytoplasm that forms a crescent around a large round nucleus.

PLASMA AND PLATELETS

Blood Is Thicker Than Water

When most people think of blood, they think of red blood cells. In fact, these cells and other formed elements of the blood make up a smaller percentage of total blood volume compared to the liquid portion of blood: the plasma. Blood also contains another component, which remains inactive until needed: platelets.

Plasma

For an average adult, the plasma makes up 55 percent of total blood volume. It is primarily composed of water but also contains dissolved gases, charged ions such as sodium and potassium, fats, carbohydrates, vitamins, minerals, and proteins.

Albumin

The most abundant materials dissolved in plasma are the plasma proteins. Albumin, which is produced in the liver, makes up the bulk of plasma proteins and serves primarily as the "stuff" in the plasma. The presence of albumin creates an imbalance that causes water to be drawn out of the tissue of the body and into the blood.

Anatomy of a Word

colloid osmotic pressure

When proteins, rather than ions, salts, or other materials, function as solutes—a substance dissolved in another substance, in this case a liquid—the water-drawing force is referred to as colloid osmotic (oncotic) pressure.

Without this vast supply of albumin, much of the fluid that leaks from your capillaries would remain in the tissues of the body and result in severe swelling (edema). Clinical edema can occur when protein production is decreased in cases of liver damage or disease.

Albumin also acts as a carrier substance. Many materials that must be transported in the blood stream are insoluble (that is, they do not dissolve) in water, which is a huge obstacle, since blood is mostly water. Because albumin is soluble in water, it can bind to and completely coat other nonsoluble materials and carry them off.

Globulins

The second most abundant group of proteins in plasma is the globulins, which are divided into three subcategories labeled alpha-, beta-, and gamma-globulins. The alpha- and beta-globulins are water-soluble proteins that function as carrier molecules, transporting water-insoluble materials such as lipids and certain vitamins. A more recognizable name for gamma-globulin is antibody. Thus, gamma-globulins function as part of the immune system to fight infectious agents in the body and provide long-lasting immunity.

Fibrinogen

Fibrinogen acts as an emergency repair protein in the event of damage to a blood vessel. In other words, it is essential in forming a blood clot and preventing blood loss. Fibrinogen is the most abundant of the clotting factors in the plasma.

Platelets

You've already heard about plasma and the formed elements of the blood such as white blood cells and red blood cells. But we can't overlook another essential component of blood: platelets. They are cellular fragments and not complete cells, and their main function is to stop bleeding.

Platelet Formation

Huge multinucleated cells called megakaryocytes located in the bone marrow shed fragments of their cytoplasm and membrane as small packages, known as platelets. In times of low platelet numbers, the body secretes the hormone thrombopoietin, which simulates the development of new megakaryocytes and more platelets.

Platelet Structure and Activation

Platelets are packages of enzymes and other materials wrapped in a membrane. They are about half the size of the RBCs. A microliter of blood can contain as many as half a million platelets. They remain inactive until a vascular injury occurs. Then they are activated, secrete their contents onto their surface, and become extremely sticky.

Endothelial cells lining the blood vessel keep the platelets inactive by secreting materials such as nitric oxide and prostacyclin. However, these cells also enrich the underlying connective tissue with a molecule called von Willebrand factor (vWF), a potent platelet activator. As long as the endothelium remains intact and contiguous, the platelets never encounter vWF and remain inactive. When an injury occurs, receptors on the platelet surface bind to vWF, which signals the rapid release of the platelet contents, causing vascular constriction (to prevent blood loss) and activating other platelets. The platelets stick to the wound site, to each other, and to RBCs and WBCs, forming a platelet plug and slowing the loss of blood from the vessel. The platelet plug is the first in a series of steps that end in the formation of a blood clot.

HEMOSTASIS

How the Body Stops the Bleeding

Several mechanisms in the body work to halt bleeding quickly in the case of a vascular injury. The process of stopping the bleeding is called hemostasis. First, the smooth muscle around damaged blood vessels will automatically constrict to restrict blood flow and limit blood loss. The platelet plug is then formed to slow blood loss. However, to stop the bleeding altogether, a clot must be formed.

Contact Activation (Intrinsic) Pathway

The activation of clotting in response to a small, localized cut is accomplished through a series of enzymatic modifications to proteins in the blood called clotting factors that are present and continually circulating in the plasma of the blood. When connective tissue materials are exposed to these plasma-clotting factors, an activation complex is organized.

The initiation of blood clotting relies on several factors, including prekallikrein and FXII (Hageman factor). When these factors bind to collagen, they convert prekallikrein to kallikrein, which activates FXII. This starts the cascade of activations with FXII activating FXI, which activates FIX. Once activated, FIX combines with FVIII, phospholipids, and calcium to form a molecular complex that activates FX, which is the first step in the common pathway for clotting.

Tissue Factor (Extrinsic) Pathway

Both initiation pathways (intrinsic and extrinsic) occur simultaneously. In the extrinsic pathway, when endothelial cells are damaged, FVII leaves the circulation and binds with TF (tissue factor), which forms an activation complex with phospholipids and calcium to activate FX and initiate the common clotting pathway.

Common Pathway

Whether activated through the intrinsic or extrinsic pathway, FX combines with FV, phospholipids, and calcium to form the prothrombinase complex. This complex converts the inactive plasma protein prothrombin into the active enzyme thrombin, which transforms fibrinogen into the active and sticky filamentous molecule fibrin. A web of fibrin begins to form and become the foundation of the blood clot as cells, platelets, molecules, and more fibrin coagulate in the site of the injury and stop the bleeding. Although the clot is now formed, it remains somewhat fragile, and the fibrin filaments must be cross-linked for stability by FXIII. It can take up to 45 minutes for the clot to fully stabilize.

Preventing clots

Removing calcium is one way to prevent blood from coagulating. Added to blood, the molecules EDTA (ethylenediaminetetraacetic acid) and citric acid inhibit clotting by binding calcium. Warfarin (Coumadin) was first identified from leeches, which use this substance to prevent blood clotting as the parasite feeds. It works by creating a vitamin K shortage at the cellular level and preventing the formation of a calcium-binding amino acid essential to clotting.

The clot does not persist forever. As the wound heals, the fibrin web retracts and assists in pulling the healthy tissue closer together. Eventually the clot is completely removed and the healing process is complete.

BLOOD DISEASES AND DISORDERS

Problems in the Blood

Since so many components of the blood are essential for so many biological activities, it is not surprising that diseases and disorders affect this system and the blood itself. Here are some common ones.

Sickle Cell Anemia

Common in populations of sub-Saharan African descent, sickle cell anemia (SCA) is a condition in which a mutation in the hemoglobin gene causes the molecule to become rigid and form crystalline structures within the RBC. These abnormal molecules give the RBCs a sickle shape and cause them to flow less efficiently through the tight passages of the capillaries.

Hemoglobin is formed from 4 protein chains; a person with SCA has a mutation in the beta chains. Only a single nucleotide in the beta chain gene is altered, but this is enough to change a glutamic acid into a valine at amino acid 6 and create the disease.

The inheritance of this gene stems from its survival in regions of malaria infection. The parasite of malaria reproduces itself within the RBC of a human host. However, a carrier of this mutation (someone with only one mutant gene who does not have the disease) will have RBCs that, when infected by the parasite, rupture prematurely, making it impossible for the parasite to reproduce. But as this gene has been passed down through generations that have not been impacted by malaria, the incidence of the condition has increased.

Anemia

This condition is most commonly associated with someone having too few RBCs (less than 40 percent hematocrit, or percentage of blood cells, in most individuals). However, a person may also be anemic if their RBCs do not contain enough hemoglobin. There are a number of causes for anemia that include insufficient iron in the diet, kidney disorders that result in lowered erythropoietin production, and abnormalities of the stem cells in the bone marrow.

Another cause of anemia, termed pernicious anemia, stems from a vitamin B_{12} deficiency. While this may be due to a diet insufficiency, pernicious anemia is an absence of a cofactor that is essential for B_{12} to be absorbed by the intestinal epithelium intrinsic factor, which is normally produced by the cells in the stomach. This cofactor usually enables the cells of the small intestine to effectively absorb dietary B_{12}. Without it, B_{12} cannot be used and RBC formation declines.

What is the difference between sickle cell anemia and thalassemia?

While sickle cell anemia is a qualitative defect in hemoglobin, thalassemia results when one or more chains of hemoglobin are not produced in sufficient amounts. Thalassemia is a type of anemia. If the problem is with the beta chain, the classification is beta-thalassemia; if there is a deficiency in the production of the alpha chain, it is alpha-thalassemia.

Hemolytic Disease of the Newborn

This condition only occurs when an Rh-negative female becomes pregnant with an Rh-positive baby. If the father is Rh negative, there is no possibility of this occurring. However, if the father is Rh positive, there is a 50 to 100 percent chance of the baby being Rh positive (depending on whether the father has 1 or 2 Rh alleles).

For a first pregnancy, there is no danger to an Rh-positive baby, since the mother does not have pre-existing anti-Rh-factor antibodies circulating in her plasma. However, during the birth process, fetal and maternal blood combine as the fetal portion of the placenta detaches from the uterus, stimulating an immune response in the mother. Antibodies produced against the Rh factor now circulate in the plasma, presenting a significant risk to the baby in any subsequent pregnancy (if that baby is also Rh positive). These antibodies cross the placenta and destroy Rh-positive RBCs in the baby, causing this disorder.

Preventing the mother from mounting an immune response against her baby's blood is an easy task. Prior to delivery, the mother is given an injection of anti-Rh-factor antibodies. These antibodies bind to any Rh factor that makes its way into the maternal blood stream. This effectively eliminates the rogue Rh factor from the blood, renders it unable to stimulate an immune response, and protects any future babies. This procedure needs to be done with each potential Rh-positive pregnancy.

LYMPH AND LYMPHATIC CIRCULATION

Bailing the Leaky Ship

The lymphatic system is another network that transports material throughout the body. While most of the plasma that leaks from capillaries into tissue is returned via those same capillaries, some fluid does not. If this interstitial fluid remained in the tissues and was allowed to further accumulate, swelling (edema) would occur. The lymphatic system is a means by which this leaked fluid is returned to the circulatory system.

Additionally, as this fluid washes through the tissues, it invariably collects cellular debris and pathogens. To identify and mount an immune response against these materials, the lymphatic system determines if the materials are pathogens.

Composition of Lymph

Once in the capillary, high blood pressure causes plasma to be pushed into the interstitial tissues, the tissues outside the vessels. This lowers the pressure of the remaining plasma in the capillary so that at the end of the capillary (before it becomes a venule) the pressure is much lower.

Plasma is rich in proteins, such as albumin, which are too large to leak into the tissues. So, while the blood pressure is lower at the venous end of the capillary, the protein content remains high and continues to exert an osmotic drawing force on the fluid in the tissues. In this way, the proteins act as solutes to attract water and this attractive force is stronger than the blood pressure trying to push out more fluid.

Therefore, the power imbalance favors fluid returning to the capillary and leaving some plasma (minus proteins) outside the capillary in the tissues. This plasma-minus-proteins is now called lymph.

Anatomy of a Word
filtration
Filtration is when fluid leaves the capillary, and adsorption is when tissue fluid returns to the capillary.

With little protein content and no pump (such as the heart) exerting pressure on it, the lymph has very little reason to move into the lymphatic circulation. It depends on pressure exerted in the interstitial tissues by surrounding organs—especially muscle. The action of walking, breathing, or any movement in general causes the position of organs to change, temporarily increasing the pressure of the lymph. This external pressure forces the stranded lymph into the lymphatic system.

Lymphatic Circulation
The lymphatic system begins with narrow, thin-walled, and blind-ended vessels that collect the lymph from tissues and funnel it back toward the heart. Along the way, these vessels release the lymph into small organs called lymph nodes, which act like a filtering system. Lymph nodes are filled with phagocytic cells (macrophages) and lymphocytes (immune cells). Within the lymph nodes, the macrophages remove and examine the debris, and if any pathogens are found, present the molecules (antigens), signifying the material as a pathogen to the local lymphocytes in order to stimulate an immune response. After leaving the lymph node, the lymph continues in even larger lymphatic vessels until the fluid, now

cleaned of debris and pathogens, is returned to the circulation via the subclavian veins.

Lymphatic Capillaries and Vessels

Capillaries are the only vessels that allow for the direct exchange of materials. So to move from the tissues into lymphatic circulation, the lymph must flow through capillaries. The ones it uses are called lymphatic capillaries. The pressure exerted upon the fluid forces the lymph between the loose and overlapping junctions of these capillaries. In essence, the junctions function as unidirectional valves to allow the lymph to come in, but not go out. Additionally, the endothelial cells of the lymphatic capillaries are attached to the peripheral connective tissue cells via fibers. These fibers help keep the lumen, or inside space, of the lymphatic capillary open as pressure pushes the fluid into the vessel.

Located within capillary beds of the circulatory system, lymphatic capillaries are in the perfect locations to collect lymph as it fails to return to the circulatory system via the systemic capillaries. Additionally, in the gastrointestinal tract, large lymphatic capillaries are present in every protrusion (villus) of the intestinal surface. Here, material absorbed from the intestines can quickly and easily make its way into the lymphatic system, and then be cleaned and screened in the lymph nodes before being passed along to the circulatory system.

Lymphatic capillaries empty their contents into larger lymphatic vessels that are composed of the same three layers (tunics) as the circulatory vessels. Lymphatic vessels most closely resemble veins with their large luminal diameter, compared to the thin wall of the vessel. Also like veins, the lymphatic vessels contain unidirectional valves that assist in the movement of the low-pressure lymph on its journey back to the heart.

How can lymph vessels be distinguished from veins?

An easy way to distinguish a lymphatic vessel from a vein is by the presence or absence of RBCs. While lymphatic vessels will have lymphocytes and white blood cells, only veins will contain red blood cells.

Large vessels combine to form lymphatic trunks. These trunks empty into one of two lymphatic ducts that transfer the drained lymph to one of the two subclavian veins, completing its return to the circulatory system.

PRIMARY AND SECONDARY LYMPHOID ORGANS

The Cleanup Crew

The lymph system uses a number of different organs to perform its functions. Primary lymphoid organs are those where lymphocytes are formed and mature. Secondary lymphoid organs are those organs that act as filters in the lymph system.

Primary Lymphoid Organs

Not all WBCs reach their final mature (differentiated) state in the marrow. Agranular lymphocytes are split between two maturation locations.

Bone Marrow

While all lymphocytes begin their lives in the bone marrow, only the B lymphocytes mature there. As naïve (immature) B cells, they are not able to recognize antigens with their receptors and are incapable of mounting an immune response. The bone marrow is compartmentalized in such a way that these immature cells are unable to enter the systemic circulation until they pass inspection, at which time they are mature and are released into circulation.

If a cell produces a B cell receptor that binds too tightly or too weakly to the antigen, then the B cell is destroyed. The most common method is apoptosis (programmed cell death or cellular suicide). B cells must also be able to distinguish between host cells and tissues and foreign pathogens before they are let out into circulation.

Name that cell

The B in "B cells" represents the maturation in the bone marrow, while the T in "T cells" is because these cells travel to the thymus to complete their maturation process.

Thymus

Located in the mid thorax (chest) in a location most often referred to as the mediastinum, the thymus is an amorphous, bilobed organ that sits superior to the aortic arch and extends upward toward the neck. Connective tissue covers the entire organ, making it an encapsulated lymphoid organ. The thymus is divided internally and is isolated immunologically from the body by cells called reticular cells. These cells provide structural support for lymphocytes and secrete hormones that stimulate their production.

Beneath the capsule of connective tissue is the cortex and below that a deeper middle region called the medulla. Different types of reticular cells are specific to these locations and isolate each region of the thymus from another. For example, some cells line the capsule and isolate the thymus from the body, while others create a boundary between the cortex and medulla. Other cells cover the capillaries and blood vessels to create a blood-thymus barrier and prevent rogue T lymphocytes from escaping into the body before they are screened.

Naïve T cells undergo a screening process similar to what occurs with B cells. In the thymus they are first screened for their antigen-binding abilities. In the medulla they are screened for their ability to distinguish between host cells and tissues and foreign pathogens. If they pass, they leave the thymus.

Once mature, lymphocytes migrate out into the body and accumulate in various locations, such as in organs of the digestive and respiratory tract, as well as stand-alone lymphoid organs. Here, they are able to quickly and effectively encounter pathogens and mount an immune response.

Secondary Lymphoid Organs
Secondary lymphoid organs are those organs that serve as filters in the lymphatic system.

Lymphoid Nodules
Often confused with lymph nodes, nodules are simply aggregates of lymphocytes. Largely composed of B lymphocytes, nodules also consist of antigen-presenting cells and reticular cells for structure and anchorage. If the nodule is a solid dark color, it is referred to as a primary nodule (follicle). This indicates that the cells of this nodule have yet to be challenged with antigen. After the antigenic challenge, the center of the nodule (germinal center) becomes lighter as the lymphocytes proliferate and become antibody-generating plasma cells. These lighter-centered nodules are called secondary nodules. T lymphocytes also reside in the nodules but to a lesser degree.

Lymph Nodes
These encapsulated, bean-shaped lymphoid organs interrupt the path of the larger lymphatic vessels. Lymphatic vessels join at the convex surface of the node. They empty lymph into the cortex of the lymph node, which is subdivided into compartments by connective tissue extensions of the capsule called trabeculae. It is within this region that the round lymphoid nodules are found.

T lymphocytes are found in the deeper or subcortical region of the node. Lymph flows through the cortex and down into the medulla (deeper middle portion) of the node, where spaces or sinuses allow the lymph to drain from the node and into lymphatic vessels.

Spleen

While the spleen does have lymphoid characteristics, it also functions in the cleaning, destruction, and removal of dead RBCs. As the splenic artery enters the spleen, progressively smaller-diameter vessels branch off and run through the bulk of the spleen. Central arteries are the smaller tributaries that are surrounded by lymphocytes, which form a periarterial lymphatic sheath (PALS). This PALS is further encompassed by a lymphoid nodule, much like that found in the lymph node. Both the PALS and the nodules make up the white pulp of the spleen.

Vessels continue to branch and radiate through the nodules until they form splenic sinuses in the area between the white pulp. This is the location of the red pulp. From these sinuses, plasma freely escapes and flows throughout the spleen, including the white pulp where antigen-presenting cells capture and display antigens to the local lymphocytes.

Tonsils

There are three sets of tonsils in the back of the mouth, all of which contribute to the lymph system's cleaning function. The three sets are:

- the adenoids
- the lingual tonsils
- the palatine tonsils

Generally when people refer to "the tonsils" they mean the palatine tonsils. These organs can become inflamed because of tonsillitis and are often surgically removed. Filled with lymphoid nodules and positioned at the boundary between the oral and pharyngeal cavities, these organs are in the perfect location to detect any pathogens trying to gain entry into the body through the oral cavity.

Also protecting the oral cavity, but to a lesser degree, are the lingual tonsils. These are located on the lateral borders of the tongue and are much smaller in mass than that of the palatine tonsils. These are filled with lymphoid nodules.

The final set of tonsils is the pharyngeal tonsils, better known as the adenoids. These are positioned higher in the pharynx at the boundary between the oral and pharyngeal cavities, and provide protection from any pathogen seeking to gain entrance to the body.

Peyer's Patches

While lymphocytes are scattered throughout the body and temporarily accumulate in areas of infection, in the final portion of the small intestine, permanent lymphoid nodules called Peyer's patches are present. They monitor bacteria growth in the intestines and prevent infection there.

Diffuse Lymphoid Tissue

Although not permanent nodules, resident lymphocytes are found in the underlying layers of both the gastrointestinal and the respiratory tracts. Classified as diffuse lymphatic tissue, or mucosa-associated lymphoid tissue (MALT), in the gastrointestinal tract it is known as the gut-associated lymphoid tissue (GALT) and in the respiratory tract it is called bronchus-associated lymphoid tissue (BALT).

INNATE AND NATURAL IMMUNITY

Your Body's Basic Defense System

The world is filled with both beneficial and harmful organisms, which only want to survive and reproduce, even if to the detriment of others, such as a host organism. Protecting yourself against macroorganisms and predators has become fairly easy. However, escaping from microorganisms is virtually impossible. Fortunately, your body has evolved a system that both prevents pathogens from gaining access to your body and destroys the pathogens that do enter before they can cause damage.

Physical and Chemical Barriers

The primary physical barrier your body uses to prevent infection is your skin. Your skin is a contiguous layer of cells and provides no access for pathogens to enter. Also, skin is composed of several cellular layers that pathogens must travel through to gain access to the deeper regions of the body and the circulatory system. If this were not a difficult enough obstacle for pathogens, new skin cells are continually added at the base of the skin and the older cells are progressively moved upward. Therefore, a pathogen's struggle to penetrate intact skin is much like a salmon's struggle to swim upstream.

The top layer of cells is shed daily, so pathogens have a limited amount of time before they are removed along with the dead skin cells. Most viruses require a living cellular host they can latch onto in order to survive and reproduce. The outer layers of skin are in fact dead cells that are compacted together into a watertight barrier. Viruses are unable to reproduce in this barrier before being shed from the surface.

While the skin is an extremely efficient pathogenic barrier, there are other means pathogens use to gain access to your internal tissues.

Anything you eat or drink may contain pathogenic agents. Although many foods are processed to destroy harmful microorganisms, the act of touching the food with your hands (especially if not washed) may contaminate the food immediately prior to eating. Many of these pathogens, which may be lucky enough to escape the lymphocytes in the tonsils, soon encounter an inhospitable environment in the stomach. The hydrochloric acid (HCl) produced to chemically digest our food denatures and destroys most pathogens that enter the stomach.

Another means of protection relies on trapping pathogens in a thick proteinaceous (protein-filled) substance called mucus. In the nasal and respiratory tract, goblet cells secrete mucus to trap pathogens, then ciliated cells move the mucus upward to the larynx where the material is transferred to the esophagus and the HCl-filled stomach. Similarly, tears are produced to moisten and lubricate the sclera of the eye. Tears also contain mucus and can trap any pathogens on the surface and remove this material from the eye via the nasolacrimal duct, which drains into the nasal cavity.

Phagocytosis and Opsonization

Many white blood cells remove pathogens and debris by phagocytosis. Macrophages are particularly adept at this. These cells remove any foreign material, regardless of what it is. Additionally, when pathogens are decorated with antibodies and/or are covered with certain markers called complement factors (discussed later), the phagocytic activity of these cells is increased. This elevation of phagocytic activity resulting in more rapid removal of pathogens is called opsonization.

Complement Factors

In an immune cascade, not unlike what happens in the clotting cascade, previously inactive complement factors, which are proteins in the blood plasma, become activated during a pathogenic attack.

The classical activation pathway is initiated via antibody opsonization of a pathogen, which triggers complement factor 1 (C1) to activate C4, then C2. These combine to enzymatically activate C3. These factors have two portions: a and b. Thus, C3 is split upon activation into C3a and C3b.

An alternative pathway can lead to the activation of C3 directly when carbohydrates are recognized. Carbohydrates are foreign to the human body, yet rather common in bacterial cells walls.

Once split, each portion of C3 is bioactive, meaning it has an effect on human tissues. C3a along with another fragment, C5a, leads to an increase in inflammation in local tissues. C3b, along with antibodies, further opsonizes pathogens. Additionally, C3b activates factors C5–C9 to form a protein pore, which inserts into the plasma membrane of pathogens and results in their death. The structure that is formed is referred to as the attack complex.

Cytokines

Many chemicals produced and secreted by immune system cells have far-reaching and powerful effects on the body. Collectively these substances are referred to as chemokines, cytokines, or lymphokines. The vast array of these chemical mediators is beyond the scope of this section; however, the general concept of their actions is essential to understanding immune function. Many of these cytokines play critical roles in the activation of the immune system during an immune response, while others are required to slow and even halt such a response.

Inflammation

Several cytokines affect the blood vessels and allow more blood to flow into the tissue (vasodilation) and to make the capillary endothelial cells easier for WBCs to migrate between (increase vascular permeability). As a consequence, more plasma leaks into the interstitial tissues more rapidly than can be removed by lymph vessels. At this point, swelling (edema) occurs. Mast cells (basophils that have left the circulation) secrete powerful inflammatory cytokines such as histamine, which makes the vessels leaky. Additionally, they secrete heparin, a molecule that inhibits the clotting cascade by preventing the activation of thrombin. If the vessels were to leak and heparin was not produced, a blood clot would rapidly form and prevent the passage of fluid and WBCs into the tissues.

Fever

Another effect of some cytokines is to increase the body's temperature above normal. A group of cytokines is called the endogenous pyrogens because of their ability to cause fever. The increased temperature helps fight an infection. Higher body temperatures destroy some pathogens directly, and reduce the effects of the toxins that bacteria can create in the body. Increased temperature also increases the division, migration, and metabolism of the immune system cells and gives them an attack advantage over many pathogens.

When is a fever dangerous?

Only when the body temperature reaches or exceeds 105°F will your cells be destroyed. However, any fever lasting for more than a few days and in excess of 101–102°F should lead one to seek clinical treatment from a doctor.

ADAPTIVE IMMUNITY

Learned Self-Defense

The specific (adaptive) immune system is divided into two components, the humoral immune system and the cellular immune system. These systems differ from the innate immune system in that your body learns to identify specific pathogens and to take steps to protect itself from those specific pathogens.

The Humoral Immune System

The humoral immune system relies on B cells that lead to the creation of antibody-producing plasma cells. As antibodies are produced, they link pathogens together into larger and larger masses, which prevents their dispersal throughout the body and provides a larger target for phagocytic cells. Additionally, antibodies opsonize pathogens for more rapid removal by WBCs.

Antibodies

When B lymphocytes are activated by their reciprocal pathogen, they begin to rapidly divide and clone themselves into either plasma cells (active antibody-producing cell) or memory B cells. Memory B cells are inactive and held in reserve for a subsequent exposure to the same antigen later in life.

Antibodies (also called immunoglobulins, abbreviated Ig) are proteins consisting of four amino acid polymers linked together via disulfide bonds between adjacent cysteine amino acids. For a general idea of the shape of an antibody, imagine a person standing with arms upward and legs together, making their body into the shape of a Y. If you split the body into right and left halves, you would have created the two heavy chains that make up an antibody. Two additional chains, the light chains, are smaller proteins attached to the arms only.

Just as you use your hands to grasp, the antibody has various regions that grasp (bind to) specific antigens. These regions are highly adaptable and changeable when needed, enabling an antibody to protect the body against an almost infinite array of antigens.

The remainder of the antibody, which would be the torso and legs, is a constant fragment. These portions are constant in the sense that when antibodies of the same isotype (genetic family) have regions that differ, the constant fragments are identical. These constant fragments, regardless of variable region or isotype, function as an opsonin for phagocytic cells.

Isotypes

While the variable regions of antibodies are specific for their antigen, the constant fragments are the same from antibody to antibody of the same isotype. The genes for the constant fragment are subdivided into different isotypes that can be used to build an antibody at certain developmental stages or for specific responses. These isotypes are arranged on the chromosome in the order they may be used. The first isotypes used are on the upstream or 5' portion of the chromosome, while the last isotype will be on the far 3' end (recall how in DNA replication, 3' or 3 prime and 5' or 5 prime indicate various points on the molecule).

The change from one isotype to another results in the splicing out of the upstream genes, meaning once you pass an isotype on the chromosome, that gene is eliminated and no longer available to be switched to later in the life of that cell or its clones.

The IgM isotype (class M immunoglobulin) is the first to be expressed on naïve B cells. This constant fragment directs five IgM antibody monomers to link together via their constant chains (the heavy chains that make up the constant fragment of the antibody), making a large pentameric molecule that is capable of binding ten antigens at a time (just as you have two hands, a single antibody has two variable regions, and can bind two antigens).

The next isotype to be expressed is the IgD isotype. This is the predominant antibody found in the circulating plasma. It is also the protein predominantly responsible for the humoral immunity. Like the IgD, IgG antibodies are monomeric in structure. These are the only antibodies that can cross the placenta during pregnancy and provide a passive immunity to the developing fetus. However, these isotypes can also cross the placenta and lead to destructive effects, including hemolytic disease of the newborn.

The next isotype is the IgA isotype, which is a dimeric antibody consisting of two monomers joined via the constant fragment. These antibodies are particularly resistant to degradation and can be found in many mucous secretions including tears, saliva, and breast milk.

The final isotype is the IgE type. This type is capable of eliciting some of the most powerful immune responses of the body. These monomeric antibodies are often linked to the surface of mast cells via an IgE constant fragment receptor. If an antigen is present, this may result in two adjacent antibodies binding simultaneously and bringing the two receptors close enough together to trigger a cellular response.

In this way, mast cells are signaled to degranulate or rapidly secrete their inflammatory cytokines. When enough of these mast cells release their chemical mediators in a mass explosion, it can trigger a severe to dangerous allergic response and send someone into anaphylactic shock.

Primary Immune Response

Upon initial exposure of the body to a particular antigen, B lymphocytes, granulocytes, and macrophages stimulate a primary immune response. The leukocytes phagocytose (consume by the process of phagocytosis, or engulfing) material and signal with cytokines, and the B lymphocytes begin to divide. When an antigen binds to the B cell receptor (which is also the antibody that a B cell will produce), the cell begins to clone itself into more B cells.

Over the course of approximately 2 weeks, the individual suffers from the symptoms of the disease as the entire immune system fights to remove the pathogen and prevent further damage. This is the natural primary immune response. However, although a few IgG antibodies result, the principal product of this response is a vast number of memory B cells. These remain in the body and remain viable for possibly the entire lifetime of the individual. With these in place, when the same antigen is encountered again, there is a much faster and more robust response.

Adapting the immune system through vaccines

Vaccines are the clinical way individuals are given the opportunity to mount an immune response, build up their memory cells, and prevent the disease. In many cases, the pathogen introduced artificially is either already dead or unable to reproduce, leading to the activation of the immune cells and not the detrimental effects of a living pathogen.

Secondary Immune Response

The purpose of the primary immune response is to create more of the specific B cells and memory cells to quickly remove the pathogen before it can cause harm. While the primary response takes days or even weeks, a subsequent exposure to the pathogen results in a mass production of IgG antibodies that flood the circulation in a matter of hours. In this manner, the pathogen is eliminated and the individual suffers no symptoms.

The Cellular Immune System

The second component of the specific (adaptive) immune system involves the T lymphocytes and requires physical contact between cells. It is therefore named cellular immunity. There are four types of T lymphocytes. Each has specific receptors capable of binding to a reciprocal receptor on cells of the body. Much like the activation of B cells leads to the production of memory B cells, stimulation of T cells leads to the formation of memory T cells.

T Lymphocytes

The first T lymphocyte to consider is the helper T cell. These cells are filled with cytokines and are responsible for the chemical stimulation and activation of the immune system. This is true chiefly of T lymphocytes and to a lesser degree B cells and leukocytes. While histologically indistinguishable from any other T cell, helper T cells display a receptor on their surface called CD4, a glycoprotein that will be used along with the complementary receptor on antigen-presenting cells to become activated when a pathogen is present. Therefore, helper T cells are often referred to as CD4 positive T cells.

Cytotoxic T cells express a different CD receptor on their surface, called CD8, which has a complementary receptor on all cells in the body except for RBCs. In this way, if any cell becomes virally infected or tumorigenic, the CD8 receptor recognizes this foreign material and activates a response. When activated, the cytotoxic T cell attacks the problematic cells with a barrage of cytokines, which potentially signal the cell to undergo apoptosis (cell death).

Regulatory T cells are also produced during an immune response. While these cells do not directly attach to pathogenic cells, or signal increases in the immune system, they do play a vital role in homeostasis.

An immune response is a form of a positive feedback loop: activated immune cells activate more immune cells. To halt this chain reaction, regulatory T cells divide until a threshold mass of cells are produced, at which time they bind to the helper and cytotoxic T cells and signal them to undergo apoptosis and stop their action. These cells, however, do not affect the memory T cells that are held in reserve.

Major Histocompatibility Complex Antigens

The complementary receptors for the CD proteins on T cells are the major histocompatibility complex (MHC) antigens. The pairings of these molecules are MHC class I, which is on every cell in the body except RBCs, and is the binding partner for the CD8 of cytotoxic T cells. If the cell is normal and healthy, the cytotoxic T cell is not activated. However, if a viral or tumor marker is displayed along with the MHCI, that leads to T cell activation and cellular destruction.

For the CD4 receptor of helper T cells, the reciprocal protein is the MHC class II that is present on antigen-presenting cells, such as macrophages and dendritic cells. These phagocytic cells destroy the pathogen, display specific antigens on their MHCII molecule, and present them to T cells in order to find the exact match for the pathogen and activate the appropriate response.

IMMUNE SYSTEM DISEASES AND DISORDERS

When the Immune System Malfunctions

Given the numbers of pathogens in the environment, the rearrangement of genes of the immune system, and the screenings that occur, it is surprising that more doesn't go wrong with the immune system. However, evolution has tested the efficiency of this highly complex system of protection for your body. Even so, immune system disorders occasionally occur. The following are some of the best known.

AIDS/HIV

Acquired immunodeficiency syndrome (AIDS) occurs when one's immune system is infected by the human immunodeficiency virus (HIV), which takes advantage of the immune system to survive and reproduce. HIV utilizes helper T cells as a host that can replicate and further infect other helper T cells. At first, the individual feels symptoms not unlike that of a common cold, which dissipate in a few weeks. Thinking the disease and danger have passed, the individual typically does not seek clinical intervention and does not display any symptoms for possibly as long as 7–10 years. This is the clinically asymptomatic period of HIV infection, when the virus is slowly increasing in numbers as the helper T cells are slowly destroyed. In the end, the immune system is rendered nonfunctional and the individual succumbs to any of a myriad of opportunistic pathogens. These late-stage symptoms are what we generally consider AIDS.

AIDS/HIV is spread mainly through sexual contact, although it can also be spread through blood transfusions, by use of contaminated needles, and from mother to child. Because the asymptomatic period can last for years, one person can infect many others without realizing it.

Allergies

Allergies are not mutations or disruptions of the immune system, but instead are instances of the immune system mounting a response to a common antigen (allergen) that stimulates a massive release of inflammatory cytokines (type I hypersensitivity). This immediate type of allergy is particularly dangerous if the allergen is inhaled or otherwise consumed into the body. The vast numbers of mast cells with their attached IgE molecules flood the system with histamine and heparin, causing a massive drop in blood pressure and systemic edema (which include the lungs and airways). This is called anaphylactic shock, and if clinical intervention isn't immediately available, death could result in a matter of minutes.

Treating anaphylactic shock

To reverse an anaphylactic response, epinephrine (adrenaline) should be administered. Anyone who suffers from severe type I hypersensitivity, especially to food allergies and insect stings, must have an EpiPen (epinephrine autoinjector) readily available at all times.

T cells are also involved in allergic responses. These are referred to as delayed (type IV) hypersensitivity. Most commonly these are observed as rashes, hives, or welts that appear on the surface of the skin as a result of contact. However, this is also the mechanism responsible for tissue transplant rejection if the tissue isn't a close enough match for the patient.

Preventing transplant rejection

The major histocompatibility complex antigens (MHC) molecules between donor and recipient are being matched at this point. Since siblings (especially identical twins) have the most similar MHC molecules, they make the best donors.

Autoimmune Disorders

Almost daily the list of human autoimmune disorders grows longer. With the numbers of cells produced, invariably and by chance, autoimmune diseases will result. A few of these better known disorders include:

- Crohn's disease of the gastrointestinal tract
- Lupus, a systemic inflammatory condition
- Hashimoto's thyroiditis, which is the most common cause of goiter in females

DIGESTIVE SYSTEM

What Goes on Down the Hatch

The human body requires fuel to power itself. This power is derived from raw material that becomes energy, and this energy is processed in the digestive system. The digestive system can be thought of as a long internal tube (alimentary canal). Food enters your mouth and is processed via mechanical and chemical digestion as it is propelled along your digestive tract. In the final portion of this tube, your intestines stop digesting the material and begin absorbing nutrients into your lymphatic and circulatory systems. The essential materials that your cells and tissues rely on for survival are then distributed throughout the body.

Mouth

The oral cavity (mouth) is defined as the space bounded by your lips (anteriorly), cheeks (laterally), and palate (superiorly). In this space, food is processed by mechanical means (teeth), chemically modified (saliva), and moved into the esophagus by the tongue. Raw materials begin this journey by first passing through the opening defined by the lips.

Teeth

Made up of the hardest substance in the body and similar in many respects to bone, the teeth are the cutting and grinding implements that begin the mechanical breakdown of ingested food. All teeth consist of a portion that rise above the gum line (gingiva), which is called the crown, and the portion below the gum line (root). Covering the crown of the tooth is a brilliant white, calcium-rich material called enamel. Unlike bone, enamel has no cells and cannot repair or produce new enamel. In time, enamel slowly wears away and is degraded by bacterial enzymes (dental caries or cavities). Beneath the enamel is the core of the tooth, which is composed of dentin. This living tissue is similar to, yet harder than, bone and surrounds the pulp cavity where blood vessels and nerves reside. Blood is transported in and out of the pulp cavity via vessels that enter and leave through an opening in the root of the tooth.

Primary Teeth

The first teeth for babies erupt through the gums starting around 6 months of age and continue erupting for the first few years of life. During this time, the first set of teeth (primary teeth) form. The first set of teeth includes 5 pairs per jaw, or 20 in total.

Secondary Teeth

The second and final set of teeth, which must last a lifetime, begin to erupt the gum at around 6 years of age. They may not finish erupting until the early 20s. Structurally, both sets of teeth are the same, although the adult teeth are larger to better facilitate chewing (mastication) in the larger adult mouth. In addition to the same 20 teeth a child's mouth has, the adult mouth contains a pair of first, second, and third molars for each jaw. The final total of secondary teeth in the adult mouth will be 32.

Tongue

The tongue plays an essential role in digestion. As the jaw muscles and teeth cut, tear, and grind the food, the tongue moves the food back and forth in the mouth (intraoral transport) to help process the food into small pieces. The tongue, along with muscular contractions in the pharynx, moves food into the esophagus.

Made up of skeletal muscle covered with a thick tough epithelium, the tongue is a versatile organ. Individual muscles are arranged into at least five different planes within the tongue, which facilitates movement in a myriad of directions.

Raised extensions of the epithelium can be found on the surface of the tongue, and come in contact with the food (dorsum of the tongue). Some of these have thickened layers of dead cells, making them very tough and abrasive (filiform papillae). Other papillae are more flattened, often with grooves on the lateral boundaries, which can accumulate material and assist in the detection of taste via taste buds embedded in the walls of the papillae.

Salivary Glands

As the teeth and tongue facilitate the mechanical digestion of ingested material, an enzymatic mixture called saliva is added through ducts from three major salivary glands. This substance begins the process of chemical digestion, as well as lubricating the material for passage down the esophagus.

Anatomy of a Word

parenchyma

The secretory portion of salivary glands is called the parenchyma (remember that the parenchyma of an organ is the functional tissue, differentiated from the supporting tissues), and typically consists of either or both serous (watery, enzymatic) or mucous (viscous, slippery) secretions. Histological identification of salivary gland tissue is largely dependent upon the percentage of each type that composes the gland.

Parotid Glands

The parotid is located on the lateral portions of the face near the angle of the jaw and the base of the ear. The parenchyma of the parotid glands is entirely serous in nature and includes enzymes such as salivary amylase that begins the breakdown of carbohydrates. These secretions are transferred to the mouth via Stensen's duct (a channel that is also called the parotid duct) and empty into the oral cavity near the second molar on each side of the maxilla (upper jaw). Although it is the largest of the salivary glands, the parotid does not produce the bulk of saliva.

Submandibular Glands

As the name implies, these glands are found just inside the lower jaws, and produce around 60 percent of all saliva. Unlike the parotid, the submandibular consists of 50:50 serous and mucous secretion-producing cells. Wharton's duct transfers the saliva from the gland to the oral cavity where it is released at swellings on either side of the frenulum (the membrane that attaches from the floor of the mouth to the underside of the tongue).

Sublingual Glands

The smallest of the salivary glands is the sublingual gland, composed almost entirely of mucus-producing cells. This material will be added to the serous secretions from the other glands and aid in the lubrication of the food before swallowing.

THE UPPER GASTROINTESTINAL TRACT

The Organs of Digestion

The upper gastrointestinal tract—that part of the digestive system (alimentary canal) from mouth to stomach—is primarily responsible for the breakdown of food. While the different regions of the alimentary canal may vary in their specific structure and function, each area has the same four-layer foundation.

Mucosa

The inner layer of the alimentary canal is the mucosal layer and is composed of the luminal epithelium, which comes in contact with the processed food. A variety of cells can be found in this layer, depending on the function of the particular region. Beneath the epithelium is the lamina propria, which has loose connective tissue and an abundance of lymphocytes. The boundary of the mucosa is a thin layer of smooth muscle called the muscularis mucosae, which functions in the mechanical processing of material.

Submucosa

This region of connective tissue underlies the mucosa and contains blood vessels, lymphatics, and nervous system plexuses that control the muscular contractions (peristalsis) of the canal. Meissner's plexus (submucosal plexus) is located in this layer and provides parasympathetic control of the various secretions in a particular region. This is part of the autonomic nervous system, particularly the enteric nervous system (the part of the intrinsic nervous system that oversees the function of the gastrointestinal system.

Anatomy of a Word

plexus

A plexus is a network of interconnecting nerves (or vessels). Plexuses (sometimes "plexi") is the plural form of plexus and refers to a group of such networks.

Muscularis

Made up of thick layers of smooth muscle, this third layer of the alimentary canal facilitates the peristaltic movement of each region. The general structure of the muscularis is an inner layer of smooth muscle and an outer layer of longitudinal muscle. Rhythmic contractions of both circular and longitudinal muscle layers moves material progressively down the alimentary canal.

These contractions are controlled by another component of the enteric nervous system, Auerbach's plexus (myenteric plexus), located between the circular and longitudinal smooth muscle layers.

Adventitia/Serosa

The outermost connective tissue of the alimentary canal is the tunica adventitia. This connective tissue allows the alimentary canal to be secured to the connective tissue of the body wall in certain areas. In other locations, the canal is not attached to the body, but is covered by a thin layer of mesothelium (squamous cell membrane) called the visceral peritoneum. In this case, the outer layer is called the serosa.

Pharynx and Esophagus

As food becomes mixed with saliva, it is processed by the teeth and tongue, and turned into a spherical mass called a bolus. This bolus is moved to the back of the mouth in preparation for swallowing. From the oral cavity, the bolus is moved into the oropharynx, which is commonly called the throat.

If the pharynx were considered a pipe, then in plumbing it would be a T intersecting pipe. The oropharynx would be the stem of the T, and would connect to the upper portion descending from the nasal cavity (nasopharynx) and the lower portion connecting with the larynx and esophagus (laryngopharynx).

As the food is swallowed (deglutition), a cartilaginous flap, the epiglottis, reflexively covers the glottis (opening of the trachea) and prevents material from being aspirated into the airway. Additionally, the epiglottis creates a ramp to help direct the bolus into the esophagus.

Composed of an epithelium that is identical to that of the oral cavity, the esophagus uses its muscularis to propel food downward and into the stomach. The upper portion of the esophagus contains a high proportion of skeletal muscle in the muscularis layer, and this can be under either voluntary or involuntary control. As the esophagus descends, the muscle transitions from skeletal to smooth. By the time the esophagus connects to the stomach, the muscle is 100 percent smooth muscle, a portion of which becomes the lower esophageal sphincter, which prevents regurgitation of stomach contents, including acid, upward into the esophagus. The esophagus passes through an opening in the diaphragm (hiatus) to connect with the stomach in the upper abdominal cavity.

Stomach

In the stomach, mechanical and chemical digestion are increased. When empty, the interior lining of the stomach contains folds called gastric rugae. These folds accommodate the expansion of the organ.

Cells

At the cardia of the stomach (the location of attachment of the esophagus to the stomach), the mucosal lining cells change from the epithelium of the esophagus, which resists friction, to an epithelium that lines the stomach and resists the harsh chemical environment. Surface lining cells (SLCs) line the luminal surface of the stomach. These cells have tight junctional complexes between adjacent cells, creating a watertight barrier to keep stomach contents from leaking into the underlying tissue. The stomach epithelium is pitted in the mucosa. These gastric pits are invaginations of the epithelium downward through the lamina propria. They populate the entire mucosal layer, vastly increasing the surface area of the epithelium and creating a protected environment in the pits for the secretory cells.

The cells in the upper or neck area of the gastric pits produce a visible mucus, rich in bicarbonate, which coats the SLCs and protects them from the harmful HCl in the stomach. Farther in these pits, parietal cells become more numerous. These are the HCl-producing (oxyntic) cells of the stomach. Populating the base of the pits are chief cells, which produce and secrete an enzyme-rich mixture and are classified as zymogenic cells. The final category of cells found in the stomach is the enteroendocrine cells, which produce a number of hormones in response to stomach activity including glucagon (to mobilize liver glycogen), gastrin (to signal HCl production), and serotonin (to stimulate stomach peristalsis).

Divisions

In the cardia of the stomach, gastric glands rich in mucus-producing cells protect against the harmful acid in the stomach. In the fundus of the stomach, the dome-shaped superior portion, and the body of the stomach, the gastric glands (fundic glands) are densely populated with parietal cells. The most inferior region, just before joining the small intestine, is the pylorus, which, like the cardiac region, is rich in mucus-producing cells that neutralize the stomach acid before transferring stomach contents to the intestine. Finally, at the junction of the stomach and intestine, the inner circular layer of the stomach is expanded into the pyloric sphincter, which regulates the passage of material from the stomach into the first portion of the small intestine (duodenum).

THE LOWER GASTROINTESTINAL TRACT

More Organs of Digestion

The lower GI tract—from small intestine to anus—is responsible for the final breakdown of food, its absorption into the body, and the elimination of waste products.

Small Intestine

The small intestine is the longest portion of the alimentary canal, approximately 20 feet long. It absorbs the raw digested materials into the blood and lymphatic system. It is divided into three regions. The first, and the shortest, is the duodenum, which is connected to the stomach. This region has an abundance of cells that produce an acid-neutralizing mucus. The jejunum is the middle region of the small intestine and is approximately equal in length to the last portion of the small intestine, the ileum.

To aid in absorption, the luminal mucosa has been expanded and folded to increase surface area. The folds or finlike protrusions are called plicae circularis and function much like the blades on the inside of a clothes dryer (which help move material around). Smaller folds called intestinal villi are present over the entire mucosal surface and are where absorptive cells are located. Additionally, apical modifications called microvilli on the lining cells maximize the absorptive area.

In addition to increasing surface area, the intestinal villi provide a space for vascular and lymphatic capillaries, allowing material to pass from the intestinal tract into the rest of the body.

Between the intestinal villi and extending downward into the lamina propria are extensions of the mucosal epithelium. These are similar to the gastric pits in the stomach and are called the crypts of Lieberkühn (intestinal crypts). Here you will find cells with functions other than absorption, including the secretion of hormones, enzymes, and acid-neutralizing mucus.

Cells

The predominant cell type found lining the intestinal tract is the surface absorptive cell (SAC). With numerous microvilli, these cells are well suited to absorb material from the intestinal lumen. With tight junctional complexes between cells, they ensure that no material in the lumen passes into the deeper tissues of the intestine.

As material moves along the alimentary canal, it becomes progressively less hydrated and, as a result, more difficult to move without causing damage to the tissues. Therefore, mucus-secreting cells (goblet cells) become more abundant in the mucosal lining.

The crypts are populated with enteroendocrine cells, which produce hormones. These cells also produce gastric inhibitory peptide (to stop the production of HCl) and cholecystokinin (causing peristaltic contractions of the gall bladder to expel bile into the alimentary canal).

The ileum (final segment) of the small intestine contains lymphoid nodules (Peyer's patches) and M (microfold) cells. There are antigen-presenting cells, much like macrophages. Large Paneth cells present in the base of the intestinal crypts are triggered by pathogens to release their secretory materials. Paneth cells produce a variety of antimicrobial enzymes and agents such as lysozyme and also release several immune system cytokines that are essential for immune system function.

Large Intestine

Although the large intestine (colon) focuses on absorption, it mostly absorbs water (approximately 1400 ml/day) and compacts the material into solid waste (feces; 100 ml/day), which is stored in the lower portion of the colon until eliminated from the body. Digested material passes from the ileum of the small intestine through a muscular sphincter (ileocolic valve) and into the beginning of the colon.

The colon does not have intestinal villi. The crypts are still present, but become shorter as the fecal material gets closer to the rectum (end of the large intestine). While the colon is not the longest portion of the alimentary canal, it stretches for approximately 5 feet and is larger in diameter than the small intestine (3 inches versus 1 inch).

Another important histological, as well as gross anatomical, difference is that the outer longitudinal muscle is only present as three bands of smooth muscle called taenia coli. These maintain a certain base level of tension on the colon and results in the folds (sacculations) of the large intestine. Additionally, spasmodic contractions help move the fecal material farther along the colon.

Cecum and Appendix

The beginning of the colon is a blind-ended pocket that is inferior to the ileocolic valve called the cecum. Material is moved upward into the first ascending portion of the colon; however, some material is invariably trapped in the cecum. Projecting off the cecum is a wormlike appendage called the appendix (vermiform appendix) where lymphoid nodules can be found in the lamina propria.

Colon

From the cecum, the ascending colon rises superiorly on the right side of the abdomen before making a 90° bend and extending across the body as the transverse (across) colon. Once on the left side of the body, the ascending colon makes another 90° turn and forms the descending colon, which extends to the lower left quadrant of the abdominal cavity. To align with the midline of the body, the colon makes an S-shaped bend as the sigmoid colon and then continues straight downward as the rectum.

Rectum

The final straight segment of the colon is the rectum, which functions in the storage of feces and its elimination (defecation). Just before the anus (external opening of the rectum), two muscular sphincters retain the material internally until voluntarily released. The internal anal sphincter is under involuntary control and is always in a state of contraction. It is triggered by fecal pressure. The external sphincter may be controlled voluntarily or involuntarily, and is made up of skeletal muscle.

Pancreas

Considered one of the accessory digestive glands, the pancreas is located in the curve of the duodenum near the pylorus of the stomach. This is a perfect location, since the exocrine secretions from the pancreas pass through the pancreatic duct, into the common bile duct, and enter the duodenum through the sphincter of Oddi.

Triggered by hormones like cholecystokinin (produced by enteroendocrine cells of the intestine) and the neurotransmitter acetylcholine (active during the rest and digest phase of the autonomic nervous system), the pancreatic cells secrete a solution of digestive enzymes as pancreatic juice, which contains enzymes that further degrade carbohydrates, proteins, and fat.

Endocrine cells are the characteristic feature of the pancreas. Aggregated into small masses, these cells form the islet of Langerhans (pancreatic islets) and are surrounded by the exocrine cells of the pancreas.

Liver

The liver, the largest gland in the body, is positioned in the superior portion of the abdominal cavity just superior to the stomach. It is divided into two major lobes (right and left) and two minor lobes (the quadrate lobe is located near the gall bladder, while the caudate lobe is near the entry of the hepatic portal vein). Blood from the digestive tract is brought to the liver via the hepatic portal vein and enters the liver at the junction of the four hepatic lobes, called the porta hepatis. This nutrient-rich blood is spread throughout the open spaces of the liver (sinusoids) and forced to make contact with the liver cells (hepatocytes), where they can metabolize this material.

The regenerating organ

The liver is the most highly regenerative tissue in the body. Two-thirds of the liver may be lost and the remainder of the liver will regenerate the portion that was removed.

Structure

Think of the liver as a high-efficiency filter. Blood is emptied into long channels between rows of liver cells (hepatic cords) and flows toward a central vein. With these rows of cells and sinusoids arranged in series, it creates what resembles a 6-sided wheel with the spokes (rows of cells) radiating toward the axle (central vein). At three or more of the peripheral points on the hexagon, there is an arrangement of a hepatic vein, a hepatic artery, and a bile duct. These always occur together at these points and are thus referred to as a hepatic triad.

As this blood travels through the hepatic sinusoids, it is in contact with the capillaries that line the spaces and separate the blood cells from the underlying liver cells. The endothelial cells of these capillaries are filled with large holes that resemble Swiss cheese. These large holes allow all material in the blood, with the exception of cells and platelets, to pour into a space (the space of Disse) between the sinusoidal capillaries and the hepatocytes. It is within this space that the material is cleaned. Resident macrophages, called Kupffer cells, are abundant in the liver sinusoids and remove any substances that may be detrimental to the body.

Functions

In addition to blood-filtering and bile-producing activities, the liver is a storage facility for carbohydrate (glycogen) and vitamins (A, D, and B). This requires the liver to process carbohydrates. It can polymerize glucose into glycogen, and conversely break down the glycogen it stores and return glucose to the blood stream.

The liver also plays an important role in protein metabolism and composition of plasma. The liver produces albumin, the most abundant plasma protein, which is critical for the maintenance of osmotic pressure in the blood stream. A by-product of red blood cell destruction and hemoglobin metabolism is bilirubin, which is further processed by the liver into a water-soluble form that can be eliminated through the urine and feces.

Fats are also metabolized and regulated by the liver, and cholesterol and lipoproteins are produced by the liver.

Lastly, the liver detoxifies the blood. It enzymatically processes toxins into a less harmful form. Ethyl alcohol is the best example of this. If you consume alcohol faster than your liver can metabolize it, your blood alcohol levels increase and coordination and judgment decrease.

NUTRITION

Fueling the Fire

Although the human body is very good at transforming materials into useable substances and storing excess energy in reserves for later use, it still relies on environmental sources for fuel and materials to maintain homeostasis.

Protein

Muscle is basically proteins organized in such a way as to facilitate work. These materials do not last for a lifetime and are in a continuous state of remodel and repair. Therefore, protein must be obtained from dietary sources to provide this critical building block for human function.

The body can transform existing materials into some amino acids (the unit of protein structure); however, some essential amino acids can only be obtained from digested food. Without a sufficient supply of dietary protein from either meats or vegetables, the human body atrophies, growth stops, and premature death can occur. A diet of proteins alone, however, does not provide a full array of essential materials for healthy body function.

Carbohydrates

Carbohydrates have been much badmouthed in the media and by weight-loss proponents. However, carbohydrates are the foundational fuel for all cellular activity in the body. Without carbohydrates, the body uses lipids and protein as survival fuel. In that case, by-products accumulate in the blood stream, resulting in a lowering of blood pH (metabolic acidosis).

That being said, not all carbohydrates are good for your health. Those carbohydrates that are broken down and flood the blood stream rapidly (high glycemic index carbohydrates) cause a spike in blood sugar, which is hormonally reduced because the excess high sugar is stored as glycogen and fat. Additionally, in a short period of time, the blood sugar rapidly drops and the individual feels weak and lethargic as the body now tries to shift metabolic gears and put stored sugar back into the blood stream.

Fats

Dietary sources of fat and cholesterol are important in the production of new cell membrane, as well as steroid hormones for the body. Excess fat consumption can lead to many health issues, including coronary artery disease, so fat intake must be carefully monitored and regulated.

Maintaining a Healthy Weight

While the composition of the foods we eat is an important aspect of nutrition and health, the *amount* of what is eaten when compared to level of physical activity is often overlooked. A healthy diet should include no more calories than are demanded by the body on a daily basis. Any amount of calories over this results in fat stores increasing in the body. An additional 500 calories consumed daily (that are not used for energy) results in the addition of 1 pound of fat over the course of a single week. Likewise, a daily reduction of 500 calories per day burns a pound of fat in a single week. Therefore, diets rich in protein, with carbohydrates of low glycemic index, and limited in fats along with careful control of the amount eaten compared to energy used will yield a healthy lifestyle that can be maintained for a lifetime.

DIGESTIVE SYSTEM DISEASES AND DISORDERS

Beyond Passing Gas

With such an extensive and diverse system as the digestive system, the possible problems, malfunctions, and diseases that can occur are numerous. Following are just a few of the digestive problems that are commonly encountered.

GERD

A growing disorder in Westernized cultures is gastroesophageal reflux disease (GERD). Basically, this is a chronic heartburn condition. Gastric acid, normally restricted to the stomach where an acid-neutralizing mucus protects the surface lining cells from damage, regurgitates past the lower esophageal sphincter and into the esophagus. Cells of the esophagus are not protected from the acid and the result is a burning sensation.

Dangers of GERD

In addition to the discomfort GERD causes, the long-term exposure of the esophageal tissue to acid may lead to esophageal cancer.

GERD is often a result of an expanded hiatus or even a hiatal hernia, where portions of the stomach project through the hiatus in the diaphragm and into the thoracic cavity. Normally, the lower esophageal sphincter is at the same level as the hiatus, and as a result the muscle of the diaphragm provides extra support for closing this important valve and keeping acid in the stomach. Repair of such a defect is often enough to resolve the GERD.

In other cases, an overproduction of acid because of diet or lifestyle can increase the occurrence of acid flowing into the esophagus. Medications including proton pump inhibitors may be prescribed to limit the stomach's production of HCl. Additionally, eating smaller, more frequent meals and not lying horizontal immediately after a meal are lifestyle alterations that can minimize the bouts of GERD.

Peptic Ulcers

While the stomach has natural means of protecting its own cells from the harmful effects of acid, in some cases damage occurs to the mucosal wall and allows the acid to penetrate into the underlying connective tissue and cause further damage. Sensitive pain receptors present in the submucosa alert the individual of pain, especially after eating a meal when the acid increases. In many cases, this wound fails to properly heal because of populations of indigenous bacteria (*H. pylori*) that collect in the wound site. Antibiotics are often effective at resolving these minor ulcerations and allow the body's wound healing process to close the mucosa.

Diarrhea

The intestinal crypt cells typically produce a secretion containing antibacterial enzymes daily (intestinal juice). When pathogens or parasites are detected in the intestine or mucosal layer, these glands shift into overdrive and produce massive amounts of fluid to flush the alimentary canal. Additionally, absorption is reduced to allow more material to flush the system in hopes of removing the problematic materials. If the diarrhea is not alleviated and if fluids cannot be retained, dehydration will occur in a short period of time, and may eventually lead to death if severe enough.

Severe cases of diarrhea

While this condition may just seem like a minor inconvenience to some, for many throughout the world it is a life-threatening condition. According to the World Health Organization (WHO), over three-quarters of a million children under the age of 5 die from diarrhea each year.

Hepatitis

Literally meaning inflammation of the liver, hepatitis has many causes, including sexually transmitted viruses (the most common cause), chronic alcohol consumption, and autoimmune diseases. During the inflammatory process, immune system cells flood the liver tissue and essentially interfere with normal liver function. Therefore, the symptoms resulting from liver damage include yellowing of the skin (jaundice), nausea, vomiting or diarrhea, and loss of appetite.

Depending on the cause of the disease, the prognosis is varied. With chronic damage to the liver, scar tissue will build up and block the regenerative ability of the liver, resulting in permanent damage and loss of function. Therefore, prevention is highly advised (this can be accomplished with a vaccine). Early administration of vaccines for hepatitis A and B yielded great success (90–100 percent efficacy) in preventing the contraction of this disease.

RESPIRATORY SYSTEM

A Breath of Fresh Air

The role of the respiratory system is rather simple: bring oxygen (O_2) to the tissues and remove carbon dioxide (CO_2) from the body. This system is divided into the conduction zone, where the air is transported but gases are not exchanged with the blood, and the respiratory zone, where gases are passed between the airways and the blood.

Respiratory Epithelium

The lining of the respiratory tract warms, humidifies, and cleans the air as it passes down the tract before entering the lungs. Specialized cells facilitate these functions. In the respiratory tract, the epithelium is arranged into respiratory mucosa, a collection of cells that give the appearance of being multilayered. However, every cell has an attachment to the basement membrane. Every cell does not, however, reach the lumen, which results in nuclei being at different levels—an example of a pseudostratified columnar epithelium.

Columnar Epithelium

The surface epithelial cells are taller than they are wide, so they are columnar in appearance. This shape is simply due to their arrangement in the layer and does not have a practical significance otherwise. However, modifications to their apical (facing the lumen or surface) membranes play a crucial role for the respiratory system.

Extensions of the cell membrane are projected into the lumen as cilia. They have a core of microtubules (a cytoskeletal polymer) that are anchored to the cytoskeleton at the base (basal body) and freely extend outward as the main body or shaft of the cilia.

What is the difference between cilia and hair?

Cilia are often compared to hairs, but are completely different in molecular organization and scale (hairs are made up of cells and cilia occur on single cells in the hundreds).

Using the microtubule associated proteins (MAPs), dynein and nexin, the core of the microtubules can be made to slide relative to each other and results in bending of the cilia. When the dynein completes the bending cycle, which requires ATP, the resulting relaxation and rebound is created by nexin, which is rather elastic in nature. This active bending and rebounding causes the cilia to wave back and forth and creates a current that is capable of moving surface material. When working in concert, all the cilia on all the columnar cells of the respiratory tract can beat to move material up and out of the respiratory tract. A sticky material that traps the particulate material aids this movement. It can then be more efficiently transported.

Goblet Cells

These are the same cells as are found in the digestive tract, which produce a thick, sticky mucus. In the respiratory system, the mucus acts as a dust trap to grab particles from the air so they can be more efficiently transported upward and eventually out of the respiratory tract.

Nose

The nasal cavity is one of the two ways air can enter the body, the other being the mouth. Air passes through the nostrils (external nares) and, because of the nasal septum, remains either on the right or left side. Once in the cavities themselves, the air first encounters the respiratory epithelium, where it begins the process of being cleared of particulates. Additionally, the air is warmed and humidified. To ensure maximum efficiency in all of these processes, the walls of the nasal cavity have several large folds that project outward into the nasal cavity and expand the surface area where air can be processed before moving farther along the respiratory pathway.

The air then moves toward the back of the nasal cavity and through narrow passages (internal nares) that lead to the top portion of the throat (pharynx). This nasopharynx connects the nasal cavities to the oropharynx, the back of the oral cavity.

Pharynx and Larynx

Commonly referred to as the throat and voice box, the pharynx and larynx respectively mark the beginnings of the respiratory tract and prevent the aspiration of liquid and solid material into the airways. This area also includes the structures essential for vocalization (sound).

Air from the nasopharynx and/or the oropharynx moves downward to the laryngopharynx. This chamber is protected from liquid and particulate matter by the epiglottis. Air easily finds its way around the epiglottis and moves into the larynx.

Within the larynx, connective tissue shelves on the lateral portions of the passageway and muscles create the adjustable cords that vibrate as air moves between them, creating sound. The most superior folds (shelves or chords) are thicker and more substantial in structure to provide a protective umbrella for the more delicate chords that are inferior. These are called the vestibular folds and they protect the vocal folds (vocal chords).

It is these delicate, thin connective tissue shelves that stretch across the laryngeal opening and become taut or loose to provide a larger or smaller opening, which changes the pitch of the sound emanating from the vibrating chords. The vocalis muscle, which is lateral to the folds in this region, directs this tension on the chords.

Trachea and Bronchial Tree

The trachea (commonly called the windpipe) marks the single passage for all air to get to and from the lungs. It lies just under the esophagus and branches off a number of times, some outside but most inside of the lung tissue itself. With each branch, the passageway becomes considerably narrower and the walls of the tubes become thinner until a dead end is reached and gases can be exchanged between the air and the blood.

The trachea is formed with intermittent cartilaginous rings that occur along its length. These incomplete, C-shaped rings provide support for the trachea, especially when you inhale, where the lower pressure would likely cause a less well-supported tube to collapse inward on itself.

The first branching of the bronchiolar tree (the branched air passages) is outside of the lung tissue (extrapulmonary). This splitting up results in a right and a left primary bronchus (plural: bronchi). Because the right lung has three lobes and the left lung has two, greater airflow is needed on the right side, so the right bronchus is larger in diameter than the left.

Primary bronchi divide and give rise to secondary bronchi within the lung tissue, marking the first intrapulmonary bronchi. The right bronchus divides into three secondary bronchi, each going to one lobe in the right lung, and the left bifurcates into two secondary bronchi for each of the lobes of the left lung.

Tertiary bronchi are the next branches that occur on the tree and provide air to a discrete and isolated section of the lung tissue called a bronchopulmonary segment. Connective tissue septa separate these regions from other segments of the lungs. They are completely independent with their own vascular supply and airway. On the right side, there are 10 segments divided unequally between the lobes of the lungs, and eight segments on the left with each lobe divided into four segments each.

These three levels of bronchi end in the next order of branching where the passageways become bronchioles. Bronchioles are much smaller in diameter than bronchi and lack the cartilaginous rings or plates of the bronchi. Layers of smooth muscle wrap around the increasingly smaller-diameter bronchioles to enable them to constrict or dilate as dictated by the autonomic nervous system. From the tertiary bronchi, there are approximately seventeen orders of branching continuing until the end of the conduction zone (the area in which air is passed through the body) is reached and the respiratory zone (where gas exchange occurs) begins.

The terminal bronchiole marks the last segment of the conduction zone (hence the name terminal). The diameters of these passageways are very thin with only one layer or two of smooth muscle present.

Lungs

The tissue that makes up these essential organs is more empty space than actual structure. Take away the bronchiolar tree, blood vessels, and nerves and what remains very much resembles an extremely porous sponge. This thin-walled and porous area marks the bulk of the respiratory zone, where gases can be exchanged between the air and the blood.

Respiratory Zone

From the terminal bronchioles, air passes to the first portion of the respiratory zone, which are called the respiratory bronchioles. These are often difficult to identify histologically because their structure is the same as the terminal bronchiole with one critical exception: the presence of alveoli. These are millions of small bubbles that mark the end of the respiratory zone for the lungs. Composed of flattened epithelial cells and having a diameter of approximately 0.5 μm (a μm is a micrometer, or one-thousandth of a meter), these structures mark the area of gas exchange.

How many alveoli exist in the lung?

There are 300–500 million alveoli in the lung. Scientists estimate that these millions of alveoli make up approximately 30–50 meters of surface area for gas exchange. This is about the size of a standard tennis court.

Following the respiratory bronchioles, the air moves into an alveolar duct. These thin-walled passages are more defined by the alveoli themselves. However, the openings of the alveoli that protrude from this duct possess a single smooth muscle cell, which creates a cellular sphincter to regulate the air moving into these areas. These are viewed as knobs on the lateral aspect of the alveolar duct.

Finally, air moves into a cluster of alveoli called the alveolar sac, which resembles a cluster of grapes. From the common atrium of the sac, air moves into any of the attached alveoli so that gas can be exchanged. Between adjacent alveoli, a small opening or pore exists to help maintain the air pressure throughout all the alveoli that are part of the alveolar sac. These are identified as alveolar pores or pores of Kohn.

INHALATION, EXHALATION, AND GAS EXCHANGE

Every Breath You Take

The active mechanism that pushes air into and out of the lungs is often termed an aspiration pump. In the case of humans, the lungs are not the pump; the thoracic cavity is.

Thoracic and Pleural Cavity

To illustrate this arrangement and describe the thoracic cavity, it is easiest to start from the outside of the body and work inward. From the outside, the first layer is the skin and muscles that form the body of the torso. Next is the rib cage, which can expand and contract to aid inhalation and exhalation.

Lining the rib cage and forming the outer boundary for the pleural cavity (potential space between two pleura of the lungs) is a thin layer of mesothelium called the parietal pleura (the parietal peritoneum of the pleural cavity). Not only does this layer of mesothelium seal the pleural cavity, it creates a low friction and nonadhesive surface that presses against the lungs.

Another layer of mesothelium, found on the surface of the lung tissue itself, is called the visceral pleura. With these two mesothelial layers adjacent to each other, intrapleural space is created. The air pressure in the intrapleural space is always slightly lower than the pressure inside the lungs (intrapulmonary pressure). Therefore, the visceral pleura is always pressed against the parietal pleura and obliterates any space between the two. This makes it a "potential space" that only exists if either pleura becomes damaged and the pressure equalizes.

Anatomy of a Word

pneumothorax

Pneumothorax is a condition in which damage to the body wall (and parietal pleura) allows the intrapleural pressure to equalize with the atmospheric pressure, causing the lungs to pull away from the pleural wall (collapse).

The pleural cavity is basically an isolated cavity that surrounds the lungs, which, under normal conditions, does not receive or lose any air. Within this cavity is the interior of the lungs, which can freely exchange air with the external environment through the bronchiolar tree. To cause air to move into the lungs, the intrapulmonary pressure must be lower than the atmospheric pressure. To cause air to move out of the lungs, the intrapulmonary pressure must be higher than the atmospheric pressure.

Inhalation

Lung tissue is not muscular and therefore cannot dilate or constrict on its own. The pump that drives ventilation is a collection of muscles in the thoracic cavity. By expanding or constricting the thoracic space, the intrapleural pressure is decreased or increased. When the thoracic cavity expands, the pressure within the intrapleural space decreases and the lungs expand. When the thoracic cavity contracts, the pressure within the intrapleural space increases and the lungs contract.

One of the main muscles involved in ventilation is the diaphragm. This dome-shaped muscle is under autonomous control (but can be voluntarily controlled) and forms the inferior boundary of the thoracic and pleural cavities. When relaxed, the superior dome portion of the diaphragm projects upward into the thoracic cavity.

When contracted, the diaphragm flattens and moves downward, which increases the volume of the thoracic cavity.

Additionally, one of the two sets of muscles between the ribs (the external intercostals) contracts. This force causes the rib cage to hinge upward, and in doing so expands the thoracic cavity laterally. Both of these actions create the lower pressure required to expand the lungs and thus aspirate air into the lungs during an inhalation.

During periods of activity when respiratory rates increase, inhalation is deeper and more rapid and requires additional muscular support. This comes in the form of various accessory muscles that are attached near the top of the rib cage and include the sternocleidomastoid, parasternal muscles, and the scalenes.

Exhalation

Compared to inhalation, exhalation is simple and passive. To reduce the size of the thoracic cavity after an inhalation, all the muscles involved in inhalation simply relax and let gravity pull the rib cage back downward. After the inhalation, the diaphragm returns to its normal dome shape and further constricts the thoracic cavity, generating high pressure inside the lungs. Therefore air is pushed outward.

As in active inhalation, active exhalation requires additional muscular assistance. The internal intercostals are arranged in a different orientation to the externals such that when they contract they assist in the more rapid lowering of the rib cage and the more forceful exhalation.

At this point in the respiration process, gases move across the respiratory membrane (by diffusion) and interact with blood elements for transportation into or out of the body.

Blood-Air Barrier

Air and blood never naturally mix, but they must come within close proximity to each other for diffusion of gases to effectively and efficiently occur. Therefore, each alveolus contains capillaries, the only blood vessels that allow the exchange of gases. The lining cells of the alveoli and the endothelial cells of the capillaries compose the thin blood-air barrier.

In the alveoli, these cells are flattened, much like the capillary endothelial cells, and are called type I pneumocytes. Another cell type present at the alveolar level is the type II pneumocytes (also called great alveolar cells). These are huge rounded cells that bulge into the lumen of the alveoli. They do not aid in the exchange of gases, but rather produce a substance called surfactant that assist in keeping the alveoli open. At 0.5 micrometers in diameter, the surface tension forces of water would be sufficient to collapse the alveoli inward upon themselves. But the phospholipid-rich surfactant interacts with the water molecules, while at the same time their hydrophobic chains keep other water molecules at a distance. In this way, the pressure required to keep the alveoli open is all but eliminated in the presence of surfactant.

External Respiration

External refers to the gases and their location. In this case, the external air has simply been inhaled into the lungs to the depth of the alveoli. However, the air is still just that, atmospheric or "external" air that must exchange gases with the blood.

The movement of gases is purely by diffusion. Therefore, the pressures of the gases in the blood versus the pressure in the air determines the direction of diffusion. Within the alveoli, O_2 has a partial pressure of approximately 105 mmHg, while that of the blood (just returning in the veins) is at its lowest at 40 mmHg. This drives the diffusion of O_2 from the alveoli and into the blood. Conversely, CO_2 pressure in the alveoli is at

40 mmHg, which is lower than that found in the blood (46 mmHg). This therefore drives CO_2 from the blood and into the alveoli, where it can be expelled from the body with the next exhalation.

Internal Respiration

During internal respiration, the gases are transferred to the tissues and cells. At the tissue level, pressure inequalities drive oxygen out of the blood and into the tissues and CO_2 returns to the blood from the cells. Although the direction is opposite from what occurred in the lungs during external respiration, the direction of movement is still from high pressure to low pressure.

Gas Transport

While some of the gases that dissolve into the blood remain in the liquid plasma, many will be bound and transported (oxygen to hemoglobin) or may be processed and transported in an alternate form (CO_2 conversion to bicarbonate ion).

Oxygen

Red blood cells are the main transportation mode for O_2 throughout the human body. Oxygen diffuses into the red blood cell and binds to an iron atom, which is held in place by the heme group of the larger protein hemoglobin. In the lungs, deoxyhemoglobin (hemoglobin lacking oxygen) binds to oxygen and becomes oxyhemoglobin so it can be transported throughout the body. Under atmospheric conditions, hemoglobin will exist in a state where 97 percent of the hemoglobin molecules are oxyhemoglobin. In fact, the love hemoglobin has for oxygen is so great that even if the partial pressure of oxygen in the air is decreased from 100 mmHg to 60 mmHg, hemoglobin remains approximately 90 percent saturated.

At internal respiration, the oxygen gradient overcomes this affinity for hemoglobin and is unloaded to the tissues from the blood by diffusion. For normal activity, approximately 20–22 percent of oxygen is unloaded to the tissues. This may at first seem like a waste, but it is in fact a reserve. Under heavy exercise conditions, up to 80 percent of oxygen may be unloaded to the tissues.

Carbon Dioxide

Although some deoxyhemoglobin can bind to CO_2 and become carbaminohemoglobin, this only accounts for a small fraction of the transported CO_2 in the blood. The majority (70 percent) of CO_2 is transported in the blood stream as bicarbonate ions that are dissolved into the plasma.

Carbon dioxide enters the blood stream and then the red blood cell via diffusion. Under conditions of high CO_2 levels, such as what exists at the tissue level during internal respiration, carbonic anhydrase facilitates the conversation of CO_2 into carbonic acid, which rapidly and spontaneously dissociates into hydrogen and bicarbonate ions. Some of the H^+ ions bind to hemoglobin, while others are transported into the plasma where they cause a decrease in blood pH. The bicarbonate ions are also transported outside of the cell in a process called the chloride shift. During this process, bicarbonate moves outward while the chloride ion is transported inward to offset the charge difference created by transporting of bicarbonate.

Once the blood returns to the lungs and during external respiration, this entire process is reversed.

RESPIRATORY DISEASES AND DISORDERS

When You Can't Catch Your Breath

Any problem that reduces the efficiency or the ease with which gases can be exchanged between the blood and the lungs will wreak havoc on all systems of the body. Sometimes disease and infections, like pneumonia and bronchitis, can temporarily cause respiratory difficulties. Other diseases are longer-term and cannot be cured.

Anatomy of a Word

hypoxia

Hypoxia is reduced oxygen levels in the blood. It leads to increased respiratory and cardiac rates in an attempt to compensate for this condition. However, this requires more energy and more oxygen, and cannot be maintained as a normal level of physiological function.

Respiratory Distress Syndrome

Respiratory distress syndrome (RDS) occurs in newborns, especially premature infants, because of an insufficient production of surfactant. Type II pneumocytes normally begin producing substantial levels of surfactant in the last few weeks of pregnancy, which is required to keep the alveoli open and functional. If babies are born too early, before surfactant levels are optimal, the alveoli are unable to remain open and much less air fills this level of the respiratory zone where the bulk of external respiration occurs.

How is RDS treated?

To treat infants with respiratory distress syndrome, exogenous (artificial) surfactant is administered into the airway and lungs of the infant in order to open as many alveoli as possible. This continues until the baby's own body is capable of making enough surfactant.

Asthma

This is a chronic obstructive disorder that makes it difficult to breathe. Triggered by either allergic or inflammatory immune mediators, the smooth muscle of the bronchioles spasmodically and involuntarily constricts and reduces the flow of air into the lungs. Wheezing and shortness of breath result, but can be reversed by blocking beta 2 adrenergic receptors in the airway, thus mimicking an autonomic nervous system response that dilates the bronchioles.

Emphysema

Emphysema is a nonreversible destruction of lung tissue, specifically alveoli. Any damage, environmental or otherwise, that causes the death of type I pneumocytes and their replacement by connective tissue cells (scar tissue) reduces the efficacy of the lung tissue. The symptoms are similar to those of other respiratory disorders and, unlike asthma, are irreversible and permanent.

Chronic Obstructive Pulmonary Disease (COPD)

COPD is an obstructive disease that makes breathing difficult. It has claimed an increasing number of lives in Western cultures as a result of smoking tobacco (smoking is the number one cause of COPD worldwide). This disease is somewhat of a combination of asthma and emphysema. It begins with lung irritation, primarily from smoke, and leads to an immune response and inflammation. Chronic exposure of the lung tissue to the harsh chemicals in the smoke results in chronic inflammation and a reflex bronchospasm response.

However, unlike asthma, these constrictions of the airway are poorly reversible and permanent in nature. If the environmental stressors are not eliminated, the reduced airflow leads to the destruction of alveolar tissue (emphysema). Together the asthma-like reduced airflow and emphysema compose the major problems involved in COPD, which is extremely difficult to treat.

Cystic Fibrosis

Cystic fibrosis is a genetic disorder that primarily affects the lungs but also affects other organs of the body, such as the pancreas and kidneys. A thick, sticky mucus forms in these organs. In the case of the lungs, the mucus makes it difficult to breathe, interferes with the effective exchange of gases, and causes lung damage. There is no cure, and treatment focuses on preventing lung infections and loosening and removing the mucus. Life expectancy for individuals with this disease has increased in recent years but it is still a life-threatening condition.

ENDOCRINE SYSTEM

How Hormones Are Produced

The endocrine system controls many of the functions of the human body, including but not limited to metabolism, reproduction, growth, and activity level.

Hormones

The chemical mediators of the endocrine system are collectively termed hormones and are distributed throughout the body via the circulatory system. The intensity of a physiological response depends on the concentration of hormones produced and the density of receptors expressed on the cell surface of the appropriate tissues. If, for example, no receptors are available to bind to even a high level of hormone, no cellular response is initiated.

What exactly is a hormone?

Any substance that is transported in the blood and that elicits a cellular response may be termed a hormone.

Amino Acid Derivatives

Individual amino acids, while usually used to create proteins, may be transformed into other biologically significant molecules such as hormones. This is the case for the amino acid tyrosine, the starting material for a group of hormones called the catecholamines. The amino acid phenylalanine is converted into tyrosine, which may be further processed into dopamine, a type of catecholamine.

Dopamine functions as either a hormone or neurotransmitter depending on where it is produced and how it is transported.

What is the function of dopamine?

In the brain, dopamine is essential for motor control, as well as for mood associated with reward and gratification. Outside of the nervous system, dopamine is a potent vasodilator and regulator of another catecholamine, norepinephrine.

In the next phase in the processing of tyrosine, dopamine is first converted into norepinephrine before it is processed into an essential body hormone, epinephrine (adrenaline). Additionally, tyrosine may be processed by thyroid cells into a differently iodinated form, either thyroxine or triiodothyronine.

Tryptophan is another amino acid that may be processed into hormones. Serotonin and melatonin are derived from tryptophan in the pineal gland, and play a role in diurnal (sleep/wake) cycles.

Proteins

The largest molecular group of hormones is the protein hormones. Many of these are produced in the pituitary gland (the master endocrine gland) and control functions from water retention in the kidneys (antidiuretic hormone, ADH), thyroid hormone secretion (thyroid-stimulating hormone, TSH), and reproduction (luteinizing hormone, LH; follicle-stimulating hormone, FSH).

Pancreatic islet cells also produce and secrete protein hormones such as insulin and glucagon, which regulate the carbohydrate level in the blood stream. Other protein hormones (calcitonin and parathormone) regulate the level of calcium in the blood stream.

Steroids

While cholesterols are often viewed in a negative light, without them the human body would not have steroid hormones. Cholesterol is the starting material upon which the steroid hormones are based. These include testosterone, estrogen, and progesterone, which function in human reproduction. Cortisol (cortisone) is another steroid hormone produced in the adrenal gland that takes part in many functions of metabolism, especially carbohydrate release into the blood stream.

Pituitary Gland

The pituitary gland (hypophysis) is the master regulatory gland of the human body. This endocrine gland produces and secretes hormones into the blood stream that lead to the release of other endocrine hormones from other glands.

The pituitary gland is found at the base of the brain and suspended by a stalk (hypophyseal stalk) just inferior to the hypothalamus. Only about the size of a pea, the pituitary is composed of two lobes that are derived from distinctly different embryological tissues.

Anterior Pituitary

The anterior pituitary (also called pars distalis and adenohypophysis) resembles glandular epithelium because of its embryonic source. The cells of the pituitary are called chromophils and include:

1. Acidophils, which are the most abundant cell type in the anterior pituitary. These cells are further subdivided based upon the hormones they secrete:
   - The somatotrophs produce the growth hormone somatotropin.

- The mammotrophs secrete prolactin. This hormone, stimulated by the pituitary hormone oxytocin, promotes mammary gland development as well as lactation.

2. Basophils are subdivided into:
   - The corticotrophs that produce adrenocorticotropic hormone (ACTH). This hormone stimulates the cortical cells in the adrenal gland to release other endocrine hormones.
   - Thyrotrophs, another basophil cell type, produce and secrete thyroid-stimulating hormone (TSH, thyrotropin). As the name of this hormone implies, it stimulates the production and release of the thyroid hormones thyroxine and triiodothyronine.
   - The final type of basophil is the gonadotrophs, which produce follicle-stimulating hormone (FSH) and luteinizing hormone (LH). In females, these hormones lead to the maturation of ovarian follicles, ovulation of a mature follicle, and lactation of the breasts at the end of pregnancy. In males, FSH stimulates the development of sperm stem cells (spermatocytes) and LH stimulates testicular cells to produce testosterone.

Posterior Pituitary

This portion of the pituitary isn't glandular at all. In fact, the pituicytes (cells of the posterior pituitary) closely resemble neuroglia cells. This tissue is derived from embryonic forebrain tissue and these cells support neurons whose axons extend from the hypothalamus to the posterior lobe of the pituitary gland.

Hormones produced in the hypothalamus are transported down the hypophyseal stalk and terminate in the posterior pituitary gland. Here they are stored in granules called Herring bodies in the axon terminals.

The hypothalamic hormones released by the posterior pituitary include antidiuretic hormone (ADH) and oxytocin. ADH (vasopressin) is released when low blood pressure is detected.

Thyroid Gland

The thyroid gland is a bilobed endocrine gland positioned just inferior and ventral to the larynx. Often, the two lobes are joined together across the midline of the trachea by a narrow strip (isthmus) of thyroid tissue. Additionally, about half of the population will have a small upward projecting lobe (pyramidal lobe) from the midline isthmus upward toward the larynx.

The thyroid tissue is organized into pools of hormones and stabilizing proteins that are stored outside the cell as colloid bounded by a sphere of thyroid cells (follicular cells). They make up the basic unit of the thyroid: the thyroid follicle. The function of the follicular cells is to produce, store, and release thyroid hormones upon the stimulation of the thyroid cells by the pituitary-derived TSH.

Follicular cells begin their production of thyroid hormone with a base glycoprotein called thyroglobulin. This protein is rich in the amino acid tyrosine. Cells will add iodine onto tyrosine.

Triiodothyronine (T_3) and thyroxine (T_4) are recovered from the colloid in the center of the thyroid follicle and function primarily to increase carbohydrate metabolism, heart rate, appetite, and respiration. At the same time, they decrease the production of cholesterol and triglycerides and aid in the reduction of body weight.

In addition to the thyroid follicular cells, other cells reside just on the periphery of the follicles. These cells are seen as aggregates of larger, more rounded cells than the cube-shaped follicular cells. These are the parafollicular cells. Parafollicular cells produce and secrete calcitonin, a hormone that inhibits bone reabsorption, thereby leaving more calcium stored in the matrix of the bone and reducing the free circulating levels of calcium in the blood stream. Parafollicular cells are often referred to as C cells.

Parathyroid Glands

Present on the posterior portion of each lobe of the thyroid gland, the parathyroid glands exist in four clusters (superior and inferior portions on both the right and left lobes). The primary cell of this gland is the chief cell. Endocrine in nature, chief cells produce parathormone (PTH), which increases blood calcium levels. It does so by stimulating bone reabsorption, preventing calcium loss in the urine and increasing vitamin D production in the kidneys.

Adrenal Glands

Best known for its production of adrenaline under stressful conditions, the adrenal gland is positioned as a cap on the superior portion of each kidney. This gland is often referred to as the suprarenal gland. The gland has an outer portion called the cortex and an inner portion called the medulla. Making up approximately 80–90 percent of the mass of the adrenal gland, the outer cortex can be divided into three distinct and functionally diverse zones:

- **Zona glomerulosa.** The most external of the cortical layers is the zona glomerulosa. Its cells are arranged in spherical structures. The principal function of these cells is the production of mineralocorticoids, such as aldosterone. Once secreted, aldosterone causes the tubules of the kidneys to absorb more sodium and secrete potassium. The result is the increased conservation of water and a reduction in urine volume.

- **Zona fasciculata.** The middle cortical layer is made up of rows of cells whose cytoplasm is filled with vesicles that give it a rather spongy appearance. These cells are called spongiocytes, and they produce glucocorticoids, such as cortisol. These hormones control general metabolism and have both anabolic (building up) and catabolic (tearing down) effects.
- **Zona reticularis.** This is the smallest layer and the one adjacent to the adrenal medulla. These cells produce androgens that produce a weakly masculinizing effect.

The medulla is populated almost exclusively by large spherical chromaffin cells and is the site of adrenaline (epinephrine) production. While this is the bulk of what is secreted by the chromaffin cells, about 15 percent of their secretion is also norepinephrine, which is necessary to convert tyrosine into adrenaline.

Pancreatic Islets

Named because they look like islands of endocrine cells surrounded by the exocrine cells of the pancreas, the cells of the pancreatic islets produce hormones that relate to metabolism, digestion, and pancreatic function:

- Alpha cells in the islets produce glucagon, a hormone secreted when low blood glucose levels are detected. This hormone results in the glycolysis of glycogen and the addition of glucose to the blood stream, thus increasing blood glucose levels.
- Beta cells are insulin-producing cells. This pancreatic endocrine hormone is antagonistic to glucagon and is secreted when high blood glucose levels are detected. In response, cells remove glucose from the blood, polymerize glucose into glycogen, and store

it intracellularly. The liver and muscle cells are well adept at this type of carbohydrate storage.

- Delta cells produce somatostatin (growth-inhibiting hormone). As the name implies, its job is to regulate the secretion of other endocrine hormones that stimulate growth, and by doing so, slow or halt the growth process in humans.
- Gamma cells are stimulated after a meal rich in protein and after fasting or exercise. These activities result in a lowering of the blood sugar level and cause the cells to release pancreatic polypeptide (PP). This hormone regulates the pancreas itself, affects glycogen stores in the liver, and stimulates the alimentary canal.
- Finally, epsilon cells secrete ghrelin when the stomach is empty so you feel the sensation of hunger.

Pineal Gland

This small pinecone-shaped gland (hence the name pineal) is located near the center of the hemispheres of the brain in the region of the epithalamus. It is involved in the production of melatonin, which is known to control sleep cycles and circadian rhythms.

ENDOCRINE SYSTEM DISEASES AND DISORDERS

When Hormones Don't Function Correctly

Because hormones control almost every function of the human body, any problem that arises in their production and distribution has severe to dire consequences. The following are a list of a few problems that occur because of issues with the endocrine system, particularly the pituitary gland.

Diseases of the endocrine system are primarily related to the overproduction of hormones, the underproduction of hormones, or tumors on the glands.

What is the most common endocrine disorder?

Diabetes mellitus is the most common endocrine disorder. Nearly 30 million people have some form of the disease.

Gigantism and Acromegaly

Both of these conditions arise from an overactive pituitary gland, or hyperpituitarism. This increase in hormone production, especially hormones involved in growth, leads to exaggerated features of the human body that may result in overall larger proportions (gigantism) or isolated enlarged body parts (acromegaly).

If the pituitary gland is active because of a congenital or developmental irregularity causing it to overproduce growth hormone throughout the lifetime of an individual, that person displays the characteristics of gigantism. However, if the pituitary becomes hyperactive after puberty, especially when the growth plates of the long bones have ossified (become bone), only those body parts capable of further growth show the increases in size. In this stage of life, the disease is termed acromegaly and displays overgrowth of body parts including the hands, feet, lower jaw, and the brow ridge of the forehead.

Graves' Disease

Just as an overactive pituitary can cause the condition of gigantism, an overactive thyroid gland causes Graves' disease. This autoimmune disease leads to the doubling in size of the thyroid (goiter), hyperthyroidism, and an increase in all of the physiological activities controlled by thyroid hormones. Individuals with Graves' disease suffer from hypertension (stemming from increased heart rate), insomnia, weight loss (due to increased metabolism), and extreme fatigue and muscle weakness. However, the most classical sign of Graves' disease is bulging eyes (exophthalmos).

Diabetes

Diabetes (diabetes mellitus) results from either a lack of insulin production or failure of insulin receptors to detect and respond to secreted insulin. Regardless of the molecular defect, the physiological result is an inability to reduce sugar levels in the blood stream. As glucose levels increase in the blood (hyperglycemia), glucose begins to be excreted in the urine (glycosuria) and eventually rises to such levels as to render the individual unconscious. Symptoms of diabetes are not unlike those seen in hyperthyroidism, which include weight loss and increased appetite. However, more frequent urination and increased urine volume distinguishes diabetes mellitus from any thyroid condition. Close monitoring of blood glucose levels, along with manual administration of insulin, are ways an individual can cope with this disease.

Diabetes Insipidus

Although termed diabetes (which is derived from the Greek word that means "siphon"), this condition has nothing to do with glucose or carbohydrate levels. Both conditions do, however, lead to the production of increased amounts of urine, hence the name diabetes. Diabetes insipidus is caused by a lack of production of ADH by the pituitary gland or failure of ADH receptors in the kidney tubules to detect and respond to ADH. Without ADH, water reabsorption cannot be regulated. This leads to the production of what many describe as copious amounts of diluted urine. Typically, an individual produces approximately 1.5 liters of urine per day. However, a person suffering from DI produces in excess of 3 liters and possibly up to 15 liters of urine per day.

Hyper- and hypothyroidism

Hyperthyroidism is the overproduction of the hormone thyroxine, which is produced in the thyroid. It leads to the inability to gain weight. Hypothyroidism, the underproduction of the hormone, leads to the inability to lose weight.

URINARY SYSTEM STRUCTURE

When You Gotta Go

The principal function of the urinary system is to remove toxins from the plasma of the blood and eliminate them from the body in the form of urine. In doing so, the kidneys filter a huge volume of plasma from the blood daily (around 180 liters of filtrate per day). However, all but approximately 1.5 liters (L) of fluid are returned to the body. The remainder becomes urine.

Depending upon the volume of urine produced versus fluid retained, the kidneys also function to regulate blood pressure by altering the fluid volume of the blood. Additionally, the kidneys reabsorb all of the essential components from the filtrate and return it to the blood stream. Among those elements are proteins and carbohydrates (glucose). All things considered, the kidneys actually are organs of conservation that also play a minor, yet critical, role in toxin elimination.

Kidneys

The bilateral kidneys are bean-shaped organs located in the lower lumbar region of the abdomen. Each receives a supply of blood from a single renal artery and a single renal vein returns the blood to the inferior vena cava. A continuous supply of blood is essential for kidney function, because urine production begins with filtered blood plasma entering the kidney tubules where the plasma is modified, concentrated, and excreted as urine.

Anatomy

The kidney has an indention (hilum) facing the midline of the body and a convex surface facing the latter part of the abdominal cavity. At the hilum, blood vessels enter and leave, as does the ureter, which is the tube that transfers urine from the kidney to the bladder. With two kidneys servicing the body, an individual may lose one kidney and the remaining organ will be sufficient for survival.

Blood Flow

The kidney is organized into an outer cortex and an inner medulla, much like those of organs such as the adrenal gland. From the renal artery, interlobar arteries branch and extend upward through the medulla until the boundary of the medulla and cortex is reached. At this point (the juxtamedullary zone), arcuate arteries branch and follow the curvature of this zone, which mimics the convex curvature of the kidney itself. Perpendicular branches from the arcuate artery course upward into the cortex as interlobular arteries. Here, branches called afferent glomerular arterioles provide blood to the capillary bundle called the glomerulus where the plasma is filtered from the blood and urine production begins.

While in a typical circulatory system venules follow capillaries, in the kidney, the efferent glomerular arteriole follows the capillaries of the glomerulus. It supplies a network of peritubular capillaries that carry out internal respiration for all the cells of the kidneys. This begins in the cortical region of the kidney as long straight extensions of vessels coursing first downward into the medulla as the arteria recta (straight arterioles) before returning upward into the cortex via the vena recta (straight venules). These two straight vessels compose the vasa recta component of the kidneys and extend peritubular capillaries between the two straight vessels, hence the conversion from arteriole to venule.

The venous blood is now drained into an interlobular vein with the remaining vessels removing blood from the kidneys mirroring their arterial counterparts (arcuate veins, interlobular veins, and the renal vein).

Renal Pyramids

The interior of the kidney is divided into units called renal pyramids. Essentially, the tubules and cells of the kidney are arranged as triangles with the apex of each pointing downward toward the hilum. In this way, as urine is produced, it is funneled to this single point where it leaves the kidney via the ureter. Each of the 10–12 pyramids end at their tip (renal papilla) and empty the urine into the first of 3 cone-shaped funnels called a minor calyx. Several minor calyces then empty into 3–5 larger major calyces that are drained by a single large funnel, the renal pelvis, which funnels urine to the ureter.

Nephrons

The network of tubes that transport and modify the plasma-derived filtrate into urine works its way through the cortex and medulla of the kidneys. At the beginning of the tubules is a capsule that surrounds the glomerular capillaries and captures the filtered plasma as it leaves the vessels and enters the tubules.

The glomerulus is the cluster of capillaries where plasma from the blood leaves and enters the initial space of a nephron. Bowman's capsule (glomerular capsule) surrounds the glomerulus to trap the filtered fluid in the space between the inner layer (visceral layer) and the outer layer (parietal layer) of the capsule. You can envision this arrangement by taking your fist and placing it in the palm of your other hand with your fingers over the closed fist. The fist is similar to the glomerulus and the covering hand and fingers represents Bowman's capsule. Together, the glomerulus and its associated capsule are termed a renal corpuscle and are only found in the cortex of the kidney.

Glomerular filtrate will flow through Bowman's space and into the beginnings of the proximal convoluted tubule (PCT) at the urinary pole of the renal corpuscle. This section of the nephron is also restricted to the cortex of the kidney much like the renal corpuscle. The cells within the PCT are well suited to start the process of reabsorbing essential materials into the blood stream from the filtrate. All glucose and all proteins, under normal physiological conditions, are reabsorbed by the time the filtrate enters the next section of the tubular network.

Additionally, 65 percent of the sodium chloride (NaCl) and water in the filtrate is actively and consistently reabsorbed in the PCT. The salt creates a hypertonic environment (greater solute concentration) that generates an osmotic pressure and draws water out of the filtrate through the PCT cells.

What is the difference between filtration and reabsorption?

Filtration is the process in which materials leave the blood stream. Absorption (reabsorption) is the process of returning materials to the blood stream.

Intermediate Loop

After the PCT, the modified filtrate begins to descend from the cortex into the medulla in the first portion of a medullary loop, termed the intermediate loop (the loop of Henle). This is called the intermediate loop because it is between the PCT and the next cortical portion of the nephron, the distal convoluted tubule (DCT). The first portion of the loop is to the PCT, and is therefore called the thick descending limb of the loop. From this region, another 20 percent of the salt and water is reabsorbed using the same mechanisms as in the PCT. By this point, of the 180 L of filtrate that is produced per day, 27 L remain in the filtrate, which may be reabsorbed under hormonal control down to the approximate 1.5 L of urine that is eliminated.

Deeper in the medulla is a much narrower section of the loop, appropriately termed the thin descending limb and, following a curve, a thin ascending limb. These regions are actively involved in water reabsorption. Following a thick ascending portion of the loop, which resembles the next tubular section histologically and partly functionally, the filtrate enters the DCT.

Like the renal corpuscle and PCT, the DCT is found in the cortex of the kidney and transports filtrate that has been processed through the earlier regions of the nephron. This region does little for the reabsorption of water. However, it does play an essential role in the acid-base balance of the urine, and in doing so affects the pH of the blood.

Having left the DCT, the last segment of the nephron, the filtrate now enters a collecting duct with a larger diameter. Several nephrons empty their filtrate into a collecting duct and begin the process of funneling the filtrate (and urine) toward the renal pelvis and ureter. As the filtrate moves down the collecting duct, the diameter slowly increases as the duct approaches a renal papilla of one of the pyramids. The ducts are now termed papillary ducts (also called ducts of Bellini) and leave the papilla and enter the space of a minor calyx via the area cribrosa of the papilla.

Is the collecting duct part of the nephron?

Although not a portion of a nephron, the collecting duct along with a nephron is considered the functional unit of the kidney and is called a uriniferous tubule.

Ureter

This muscular tube (3–4 mm in diameter) connects the kidney with the bladder. Much like in the alimentary canal, the smooth muscle of the ureter spasmodically contracts in peristaltic movements to propel urine toward and into the bladder. This is typically the location where one becomes aware of any kidney stones that are being passed through the urinary tract. Although the lining of the ureter is transitional epithelium, which can stretch as the tract fills with urine, it cannot expand to the extent that most stones need to pass freely.

Bladder

The urinary bladder receives urine from each kidney via the ureters and can store approximately 500 ml to 1 L of urine. Lining the bladder is the same compliant transitional epithelium found in the ureter, as well as underlying layers of smooth muscle that collectively are referred to as the detrusor muscle. These layers are under autonomic nervous system control, and when stretched reflexively contract in order to urinate (micturition).

At the inferior apex of the bladder is the opening of the urethra, the tube that transfers urine from the bladder to the outside of the body. The internal urethral sphincter, which prevents urine from leaking into the urethra, is found here. As the urethra passes through the perineal muscle of the lower pelvis, the skeletal muscle creates the external urethral sphincter.

Urethra

Much shorter in females than in males, this conduit for urine has much less smooth muscle than the ureter, and shifts from transitional epithelium near the bladder to stratified squamous epithelium (similar to the skin) near the external urethral orifice (opening at the end of the urethra).

URINARY SYSTEM FUNCTIONS, DISEASES, AND DISORDERS

The Skinny on Urine Production and Problems with It

Urine production begins when plasma leaves the glomerulus and enters Bowman's space. Throughout the nephron and the collecting duct, materials are removed from the filtrate (reabsorption) or added to the filtrate via the tubule cells themselves (secretion) in order to conserve essential materials and eliminate those in excess, and to balance blood levels and remove toxins.

Urine Production

Urine production occurs as the result of the filtration process blood plasma goes through in the urinary system. A number of special vessels are required for this process to function.

What is glomerular filtrate?

Glomerular filtrate is often termed ultrafiltrate since it is not exactly the same as the plasma found in the blood. In fact, plasma passes through three filtration mechanisms before it is able to enter into Bowman's space.

Glomerular Capillaries

Glomerular capillaries are a fenestrated type of capillary. This means that rather than materials diffusing first into the cell then diffusing out the other side, materials may pass through the pore in the endothelial cell for easier exiting of the blood stream. These pores are approximately 70–90 nm (nanometers) in diameter. Plasma is able to freely flow through these fenestrae with only the formed elements such as RBCs, WBCs, and platelets being retained in the blood vessels.

Basal Lamina

Underlying every epithelial cell is a molecular-rich layer called the basal lamina. The basal lamina is divided into two regions. Adjacent to the basal membrane of the cell is a less dense portion termed the lamina rara. An abundance of adhesive glycoproteins such as laminin and fibronectin provide anchorage points for the cells to attach to this extracellular foundation. Also found in the lamina rara are heparan sulfate proteoglycans. These consist of long linear protein cores. Polymers of disaccharides called glycosaminoglycans attach to these cores. This complexity is essentially geared to localize a high concentration of sulfate groups (negatively charged) in this layer to trap positively charged material from the plasma as it is becoming filtrate. Therefore, the lamina rara presents an ionic filtration step in filtrate formation. Also, since this layer is adjacent to the glomerulus, it is more specifically termed the lamina rara interna.

Below this layer is the much thicker and darker lamina densa. The greater density of this layer compared to the rara is due to the presence of type IV collagen, which is arranged into a meshwork much like a fishing net. Woven together, the spaces between opposing collagen molecules restrict passage of materials to anything smaller than 69 kDa (kilodaltons) and form a size-filtration step. The third layer, another lamina rara, is below the lamina densa, but is adjacent to the outer epithelial layer. This is called the lamina rara externa. It is composed of the same components as the lamina rara interna.

What is a kilodalton?

A kilodalton is 1,000 daltons. A dalton is a measurement on an atomic or molecular (that is, very tiny) scale. One hydrogen atom has a mass of 1 dalton.

Visceral Layer of Bowman's Capsule

The cells of this layer are tightly adhered to every loop of the glomerular capillaries and create a functional barrier (filter) for the filtrate to pass through. The modified epithelial cells that compose this layer are called podocytes (*pod* meaning "foot") because they have extensive and intricate cellular processes that completely cover the capillary loops. Podocytes interlock similarly to putting your hands together and interlocking your fingers to cup your hands to drink water. In doing so, spaces still exist between the fingers from which water can leak. These filtration slits create a portion of the third filtration step in the renal corpuscle. Between these slits, a membrane extends to further regulate the passage of materials into Bowman's space.

Countercurrent Multiplier

In the intermediate loop, a continuous cycle of salt reabsorption followed by the osmosis of water out of the filtrate is created in the medulla of the kidney. Water is reabsorbed at the level of the descending limb, primarily due to the active pumping of Na (and the passive diffusion of Cl-) that happens in the ascending limb.

Sodium-potassium pumps actively transport Na^+ into the interstitium (tissue surrounding the loop) and K^+ is pumped (secreted) into the filtrate to offset the loss of Na^+. Chloride follows the electrostatic attraction of Na^+ and diffuses into the medullary interstitium to combine with Na⁺. It is important to note that the ascending limb is *not* permeable to water, so the filtrate becomes less concentrated with salt, which makes it hypotonic.

Because the NaCl from the ascending loop is accumulating in the interstitium, the deeper levels of the medulla, as well as the filtrate in the descending limb, become more hypertonic (1200–1400 mOsm: an mOsm being a milliosmole, or 1/100th of an osmole, a unit of measurement of osmotic pressure). There are no additional Na⁺/K⁺ pumps in this region of the loop; however, the descending limb is permeable to water, allowing water to be reabsorbed into the interstitium. This causes the filtrate to become more and more hypertonic (1400 mOsm at the bottom of the loop).

As salt and water are reabsorbed into the medulla of the kidney, the vasa recta and peritubular capillaries are responsible for removing the excess salt and water and returning it to the blood stream. If this material were not removed, the kidneys would reach too high a level of salt concentration and be unable to increase further, which would shut them down. Therefore it is as essential to remove the salt and water from the kidney as it is to remove these materials from the filtrate so that the cycle can continue.

Collecting Ducts

Due to the active (and passive) movements of solutes into the medullary interstitium and the removal of water by the vasa recta, the filtrate in the collecting duct is more dilute than even plasma (i.e., hypotonic). Remember that the collecting ducts run through the medulla on their way to the calyces and will encounter the same hypertonic, salt-rich environment, which drew water from the intermediate loop.

At this point, the final 27 L of the original filtrate is under hormonal control, primarily from antidiuretic hormone (ADH). Water is able to leave the collecting duct through aquaporins (protein channels that specifically allow the passive movement of water) in the plasma membrane of the collecting duct cells (principal cells). ADH leads to increased numbers of aquaporins in the membranes, and therefore greater reabsorption of water. However, even during dehydration, ADH cannot lead to more highly concentrated urine (above that of the interstitial fluid).

What is the minimum amount of urine produced per day?

The kidneys produce a minimum of 400 ml of urine per day even in the face of severe dehydration. This is called the obligatory water loss. This volume is required to remove the wastes from the blood.

Macula Densa

As the ascending loop returns from the medulla to the cortex, it transitions into the DCT and passes immediately by the vascular pole (entry and exit point of the glomerular arterioles) of its own renal corpuscle. In this region, next to the glomerulus, cells of the DCT are compressed together into the macula densa (literally translated "dark spot"). These specialized cells monitor the concentration of sodium and chloride and in doing so indirectly monitor blood pressure. For example, decreased blood pressure results in a decreased concentration of sodium and chloride ions at the macula densa. This is due to reduced filtration by the glomerulus. In response, the macula densa cells release prostaglandins, which trigger granular juxtaglomerular (JG) cells lining the afferent arterioles to release the enzyme renin into the blood stream, which in turn stimulates the production of chemicals that create thirst and cause blood vessels to constrict, increasing blood pressure. This process of regulating blood pressure and water balance is called the renin-angiotensin-aldosterone system (RAAS).

Along with the macula densa, the juxtaglomerular (JG or granular) cells contribute to what is called the juxtaglomerular apparatus.

Electrolyte Balance

As with the homeostatic balance of any system in the human body, changes in one area or the concentration of one element may have drastic effects on another. Therefore, as sodium and chloride are reabsorbed (to conserve water) and potassium is secreted to offset the conservation of sodium, an imbalance in electrolytes can easily and quickly occur if not regulated closely.

Acid-Base Balance

Since H^+ can be secreted into the urine, the kidneys play an important role in the maintenance of proper blood, and therefore body, pH. H^+ ions are filtered through the glomeruli and may be secreted into the filtrate via an antiport mechanism with Na^+ in the DCT. This could explain the slightly acidic nature of urine.

Diseases and Disorders

Any condition that reduces the efficacy of the urinary system can have drastic effects on blood pressure (due to unregulated blood volume) and body toxicity (due to retention of wastes), and may eventually lead to kidney failure and death.

Kidney Stones

If the kidneys concentrate the urine excessively, then some minerals become supersaturated and yield a greater chance that through a nucleation event, they begin to form a crystal. As a seed crystal forms, the minerals accumulate on the crystal and increase in diameter. These kidney stones, also called renal calculi, can be passed in the urine if the crystal is less than 3 mm in size (the average diameter of the ureter). In this case, the individual may not even notice the stone. However, larger stones, which have jagged edges, can become lodged in the ureter until the building pressure of the urine forces the stone down the ureter. This causes considerable pain as well as damage. Often blood from the ureter appears in the urine (hematuria). If a stone becomes too large to pass safely, sound waves can be used to bombard the stone and shatter it into small enough pieces to pass through the urinary tract. This technique is termed lithotripsy (*litho* is derived from the Greek word meaning "stone"). Other techniques to fragment the stone into smaller pieces include laser catheterization to focus the laser beam directly on the stone.

Nephritis

Nephritis is inflammation of the kidneys. This is often caused by an infection. Other causes may include increased exposure of the kidneys to toxic agents or an autoimmune reaction that reduces kidney function. Typically, the patient exhibits reduced urine production, possibly blood in the urine (due to damaged renal corpuscles), and increasing levels of nitrogenous wastes accumulating in the blood (uremia). While infections of the body are not uncommon and most are not thought to be life threatening, nephritis is a serious disease and is one of the eight leading causes of human death worldwide.

Urinary Tract Infection

Any infection of the urinary tract may be termed urinary tract infection (UTI). Infections in the upper urinary tract are much more serious than those of the lower tract. Most commonly caused by the bacteria of the alimentary canal (*E. coli*), the infection begins in the urethra and, as the bacteria proliferates, extends upward in the urinary tract. Symptoms of a UTI include but are not limited to:

- painful urination (dysuria)
- cloudy urine due to the excessive number of bacteria populating the urine
- an increase in the sensation of urgency
- an increase in frequency of urination

Antibiotics and increased fluid intake (to cause increase urine production to flush the bacteria from the urinary tract) are the most common modes of treatment for a lower UTI.

MALE REPRODUCTIVE SYSTEM

Making Babies, Part 1

The true goal for any organism is to survive long enough to procreate and ensure the continuation of the species. In order for human beings to reproduce, a spermatozoon must be introduced into the female reproductive tract and fertilize an egg. This creates new life.

Testes

Similar in function to the female ovaries, the testes are where sperm form. It is also where the genital ducts begin.

Anatomy

Suspended from the perineum of the male pelvis, the bilateral testes are covered in skin called the scrotum. The scrotum functions as more than just a case for the testes; the scrotum and its thin underlying muscles regulate the temperature of the testes by either retracting and pulling the testes closer to the pelvis (to warm the testes) or relaxing and allowing the testes to descend farther away from the body (to lower the temperature of the testes).

Beneath the skin of the scrotum is a dense connective tissue capsule that surrounds each testis called the tunica albuginea. This connective tissue also accumulates at the posterior portion of the testes to form the mediastinum testis. From this point connective tissue septa emanate throughout the testis to divide the tissue into lobes. Each divided lobe contains 1–4 seminiferous tubules, which are the sites of sperm germ cells as well as where the sperm develops.

Cells

In addition to the germ cells, two other cell types are present in the testes that play essential roles in spermatozoa generation as well as the maintenance of secondary male sexual characteristics, which includes greater body hair, heavier bone structure and muscle mass, lower body fat percentage when compared to females, and the development of male genitalia.

Sertoli cells (nurse cells) are located throughout the seminiferous tubules and function to sustain and protect the developing sperm. As sperm germ cells undergo meiotic cell divisions and genetic recombination, the resulting sperm are genetically and immunologically different from the male in which they are produced. If the Sertoli cells did not create a blood-testis barrier, the newly formed spermatozoa would elicit an immune response and be destroyed. Additionally, these cells produce a fructose-rich secretion that nourish the sperm in their protected environment within the lumen of the seminiferous tubule.

Outside of the seminiferous tubules, in the interstitium of the testis, are the interstitial cells of Leydig. These cells are endocrine in nature and produce the androgen testosterone.

Intratesticular Tract

The pathway sperm use to make their way from the testes to the female reproductive tract is collectively called the male genital ducts or reproductive tract. After the sperm are formed in the seminiferous tubules, they move toward the mediastinum testis and straight, narrow terminal portions of the seminiferous tubules, which are called the tubuli recti (literally translated "straight tubes").

These short, straight passages allow the spermatozoa to move into an anastomotic (interconnecting) maze of passages in the mediastinum testis (rete testis). These allow the sperm to move into the next portion of the genital duct system, the efferent ductules. There are 10–20 of these ductules that transfer spermatozoa from the testes into the first portion of the extratesticular tract, the epididymis.

Spermatogenesis

Spermatogenesis is the umbrella term that encompasses all of the cellular and molecular processes that change male germ cells (spermatogonia) into free, mobile sperm.

Spermatocytogenesis

During this phase, the spermatogonia divide in order to reproduce themselves as well as to produce the next stage of cells in the developmental process, those being the primary spermatocytes. While the spermatogonia are situated at the basal region of the seminiferous tubules between adjacent Sertoli cells, the primary spermatocytes migrate toward the lumen and through the tight junctional complexes that separate the lumen and the tubule from the rest of the male testes.

So far, the cell divisions have been accomplished using mitosis or cloning of the cells. In later stages, meiosis must occur to produce cells with half of the genetic material (haploid or 1N cells). The main cells visible in the seminiferous tubule are the spermatogonia (with its dark nucleus in the basal area) and the primary spermatocyte (with its larger, vesiculated nucleus, due to condensing chromatin). The next cellular intermediate is formed by the first meiotic chromosomes dividing into 2 haploid secondary spermatocytes. These 2 cells rapidly divide during the second meiotic division into 4 spermatids.

Spermatidogenesis

This process of division into spermatids (called spermatidogenesis) is so rapid that secondary spermatocytes are not typically visible in a histological preparation. Although meiosis has occurred to transform a single primary spermatocyte into 4 early spermatids, only the nuclei have divided. The cytoplasm of these 4 daughter cells remains attached to one another via cytoplasmic bridges.

Spermiogenesis

This final stage in sperm formation converts the rounded early spermatids into late spermatids, which are then separated from the other daughter cells and released into the lumen of the seminiferous tubule as spermatozoa. Four distinct stages occur to shape and form all of the cellular and molecular components of the spermatozoa prior to their undocking from the Sertoli cell.

The first phase of spermiogenesis is called the Golgi phase, during which enzymes are released and provide a means by which the head of the sperm may fuse with and be inserted into the egg. While this is happening at the head of the sperm, at the opposite side, centrioles are forming into a microtubule-organizing unit to produce the base of the flagellum (the axoneme), which also course through the core of the flagellum itself.

The cap phase is next. This stage is marked by the movement of all the granules to just above the nucleus of the spermatid, forming the acrosome (acrosomal cap). Mitochondria are also moved into the area of the flagellum base to provide energy that drives the movement of the flagellum and propel the spermatozoa.

In the third stage, tail formation, the microtubules extend to push the plasma membrane outward and form the elongated structure that provides mobility to the spermatozoa. The spermatids at this point are oriented so that their now tapered heads are directed to the lumen of the seminiferous tubule.

During the final maturation phase, the excess cytoplasm is shed and the spermatids are freed from their sibling cells into independent cells. Any other remaining cytoplasm is removed while the spermatozoa are in the lumen of the seminiferous tubule by a process called spermiation (sperm release). However, at this point, spermatozoa are immobile and remain incapable of fertilizing an egg.

Extratesticular Reproductive Tract

Once the spermatozoa have left the efferent ductules, they enter the extratesticular ducts. The first portion of these ducts is found within the scrotum. The later portion is part of the spermatic cord that rises to the pelvis and enters the lower abdomen to join with the urethra and exit the body.

Epididymis

The first portion of the extratesticular duct is the epididymis, which stores sperm and reabsorbs fluid. This highly coiled tube possesses a head, which receives spermatozoa from the efferent ductules. From the superior portion of the testis, the body of the epididymis extends downward before forming the tail at the inferior portion of the testis. This is primarily where the spermatozoa are stored for possibly up to 2-3 months. The epithelial cells of the epididymis are also well suited to assist in the reabsorption of materials in the reproductive tract. Extending into the lumen are long cellular processes called stereocilia. Stereocilia do not assist in movement of materials; in fact, these processes are actually extremely long microvilli that function by increasing the cellular surface area for maximal reabsorption of material.

Vas Deferens

Beginning at the tail of the epididymis, the vas deferens extends upward from the scrotum in the spermatic cord and enters the body from an opening in the lower pelvis. This is the thickest of the ducts in the male reproductive tract because of its extremely thick layers of smooth muscle. These become active during ejaculation and generate peristaltic contractions that propel spermatozoa along and out of the male reproductive tract.

Ejaculatory Duct

As the vas deferens approaches the urethra (the common duct for urine and sperm), it is joined by a duct from the accessory sex gland known as the seminal vesicle. When this union occurs, this terminal portion of the vas deferens is renamed the ejaculatory duct.

Urethra

The final duct of the male reproductive tract is the urethra, which expels urine. Urine is harmful to sperm; therefore, several accessory sex glands are located throughout this portion of the male reproductive tract to not only nourish but also protect the sperm.

Glands of the Male Reproductive Tract

Since sperm are essentially a nucleus with half a complement of chromosomes, mitochondria (to provide energy for flagellar movement), and a bag (acrosome) of enzymes to be used to penetrate the oocyte (female egg), they have no means of generating or processing materials to use as fuel for the mitochondria. They instead must obtain these materials from the secretions of the accessory sex glands. Additionally, for protection against the acidic urine that may remain in the urethra, other secretions will neutralize and buffer the urethral environment and protect and feed the spermatozoa that are nearing the end of their journey through the male reproductive tract.

Seminal Vesicles

These accessory glands provide the majority of fluid in the total volume of semen (approximately 70 percent is produced here). This secretory material contains a fructose-rich fluid that will be used by the sperm to power their propulsion through the remainder of the male reproductive tract and freely swim through the female reproductive tract.

Prostate Gland

Situated around the first portion of the urethra, the prostate is about the size of a walnut. This gland secretes materials into the prostatic urethra that are primarily alkaline in nature and reduce the acidity of both the male and female reproductive tract to lengthen the viability of the sperm. In addition, the prostate contributes the bulk of the remaining volume of semen.

Bulbourethral Glands

Located at the base of the penis and around the membranous/ spongy urethral boundary, the bulbourethral glands (Cowper's glands) produce a lubricating fluid that is excreted at the initiation of an orgasm and precedes the semen out of the penis during an ejaculation.

External Genitalia

During embryonic and early fetal development, the gonads (ovaries and testes) and the external genitalia in males and females are indistinguishable. The same developmental tissue is used to make either male or female external genitalia. For females, the embryonic tissue remains open and does not fuse along the midline. However, in male development, under the control of male hormones, this embryonic tissue closes up into the scrotum and the penis.

Scrotum

As mentioned previously, this is the compartment of skin that encases the testicles and regulates the temperature of the male gonads. Along the midline of the scrotum, between the two testicles, is a raised line running the length of the scrotum called the scrotal raphe, which is the seam where the embryonic tissue fused together to form a sack (into which the testicles descend). Also during early development, the gonads begin to form in a similar manner in the same abdominal location. When the process of shaping the embryonic tissue into a testis is nearing completion, the testes descend through an inguinal opening and drop into the scrotum.

Penis

The male sex organ, the penis, consists of four tubes, wrapped within a tube or connective tissue surrounded by skin. Three of the inner tubes are spongy columns of tissue that can be rapidly filled with blood during arousal to generate an erection. The veins that drain these spongy columns of erectile tissue close during an erection and open following an ejaculation and orgasm. The two larger columns, which are arranged side by side on the dorsal aspect of the penis, are the corpus cavernosa. The smaller column is located just ventral to the junction of the two larger columns and will also contain the penile urethra.

The end of the penis is enlarged into the glans penis (head of the penis). At birth, this is covered by an extension of skin (foreskin) from the shaft of the penis. The penis can expand outward from this extension of skin during an erection.

What is circumcision?

Circumcision is the surgical removal of the flap of skin (foreskin) at the head of the penis, frequently done shortly after birth, for social, religious, or aesthetic reasons. Little evidence has been demonstrated to show a health benefit either way.

Male Sex Hormones

Male hormones that must shift the indifferent gonad and genitalia toward a male pathway and that must maintain that physiology are critical for normal male development and function. Early in embryonic development, testis-determining factor (TDF) hormone is produced that will initiate a cascade of molecular and cellular switches, resulting in male development.

One important function of TDF is to initiate the expression of testosterone by the interstitial cells of Leydig, which continue male sexual development. This hormone is used to maintain these activities in the adult. Low testosterone levels (Low T) result in energy loss, increased body fat deposition, and potential erectile dysfunction and/or infertility.

However, with the indifferent embryonic tissue, it isn't sufficient to turn on the male genes. The tissue that would have been used to form the female reproductive tract must be inactivated. This is also a cellular event triggered by TDF. The cells that facilitate this female "off" signal are the Sertoli cells. These cells make a hormone that shuts off the process for female reproductive development. Considering how this "off" signal is essential, logic dictates that the default pathway for gender development in humans is femaleness (unless the Y chromosome is present to shut down this pathway and turn on the male developmental mechanisms).

MALE REPRODUCTIVE SYSTEM DISEASES AND DISORDERS

Problems Relating to Sex and Reproduction

Any problem with the male reproductive tract and/or genitalia can present physiological as well as social and emotional difficulties for the individual. Therefore, with any condition involving the ability of an individual to procreate, health professionals must also be concerned about the mental anguish and social stigma the individual may presume and/or encounter.

Cryptorchidism

Failure of the testes to descend from the abdomen into the scrotum is termed cryptorchidism. While this can be quite alarming for new parents, in most cases it spontaneously resolves within the first few months of life. However, if not, this can be remedied by a surgical procedure called an orchiopexy.

Testicular Torsion

The surgical procedure just mentioned is also used to resolve the condition of testicular torsion. As explained earlier, the vas deferens as well as arteries, veins, and lymphatic vessels make up the spermatic cord. If the testis rotates within the scrotum, these vessels may become strangled as the vas deferens becomes twisted and braided within the cord. If not resolved, the reduced blood flow to and from the testis results in tissue death of the testis.

Male Infertility

The leading causes of male infertility are low sperm count and low sperm motility (movement). It isn't enough to simply have sperm in the semen; there must be sufficient numbers so that those spermatozoa arrive at the end of the female reproductive tract in sufficient numbers for one to fertilize the egg. Many are required to rupture through the outer layers that protect the egg so that one (and only one) may fertilize. Additionally, if the sperm are present in sufficient numbers but have a reduced motility, their numbers are reduced at the time required for fertilization.

Why might sperm have reduced motility?

Defects in sperm shape or design reduces their ability to move. Once such common defect are sperm that possess two tails. Their wavelike movements don't allow the sperm to move effectively.

Erectile Dysfunction

Erectile dysfunction (ED) is a source of much research and has resulted in many drugs and treatments that can aid in this difficulty. As men age, the effectiveness of the veins that must close and trap the blood in the erectile columns in the penis become less effective and the resulting erection is insufficient for sexual activity.

Prostate Cancer

Cancer of the prostate gland is often thought to be the male homolog for breast cancer in females. As men age, the prostate increases in size normally and can cause urinary and possibly sexual dysfunction. However, the more rapid enlargement of the prostate could be due to the rapid division of prostate cancer cells. Most commonly this normally slow-growing cancer occurs in men over the age of 50. Therefore, it is recommended that men over the age of 40 begin to include digital rectal exams as a part of the annual checkups.

How effective is the PSA blood test for detecting cancer?

A blood test for prostate-specific antigen (PSA) could alert the physician that the patient has an increase in prostatic tissue, possibly from cancer. This test, however, has been met with mixed data that possibly shows that detection by PSA doesn't increase the life expectancy of patients. Certainly, genetics has an impact on the risk of a person developing cancer. If a father or uncle or even grandfather had prostate cancer, the risks of developing this form of cancer increase.

FEMALE REPRODUCTIVE SYSTEM

Making Babies, Part 2

Several developmental biologists have suggested that a human is simply an egg's way of making another egg. In this regard, the female reproductive system is the means by which an egg is made. It is also the location where the sperm fertilizes the egg, and where the fertilized egg divides, matures, and becomes an independent, living individual.

Ovaries

Ovaries are the organs for the storage, development, maturation, and eventual release of the egg. These paired organs are found in the lower abdominal quadrants on the right and left sides and are attached to the pelvis via ligaments.

Anatomy

The paired almond-shaped ovaries are divided into an outer cortex and inner medulla and are covered by a connective tissue capsule called the tunica albuginea.

Within the cortex, there is a connective tissue network of stromal cells and ovarian follicles in various stages of development. Each of the ovarian follicles contain an egg precursor (oogonia) as well as the supportive cells (follicular cells) that surround the future egg.

The medulla of the ovary is populated with larger maturing follicles and the remnants of follicles from previous cycles that are in various stages of degeneration and regression.

Follicles

Follicles are the basic units of the egg (oocyte) and supportive cells that surround the egg and occur in various developmental stages within the ovary.

The most abundant of the follicles and those found at the periphery of the cortex, adjacent to the capsule, are the primordial follicles. These are made up of a single layer of flattened follicular cells surrounding a primary oocyte (egg). As the smallest of the follicles (approximately 25 mm), the oocyte is paused in prophase I of meiosis, a process that continues when stimulated by hormones and leads to the maturation of the oocyte.

The next stage of follicular development is the unilaminar primary follicle. As the name implies, this is an oocyte that has grown much larger (approximately 100–150 mm), and is surrounded by a single layer of cube-shaped cells, which is the principal characteristic of this stage compared to other follicular stages. Throughout follicular development, these cells may often be called granulosa cells. Another difference between primordial and primary follicles is the formation of a molecular zone situated between the oocyte and the granulosa cells called the zona pellucida. This layer remains associated with the egg even after ovulation and presents a molecular barrier that the sperm must penetrate in order to fertilize the egg.

Still a primary follicle, but with multiple granulosa layers, a multilaminar primary follicle marks the next stage of follicle maturation.

The granulosa cells begin to produce a fluid that is secreted into the spaces between the cells and pool into fluid-filled spaces within the layer. These small pools come together as more fluid is produced until a single large cavity is produced called the antrum. This fluid, the liquor folliculi, contains hormones such as progesterone, and is the driving force behind the follicle's expansion until it is the size of the entire ovary. This leads to a hydrostatic pressure that ruptures the follicle and the capsule of the ovary (ovulation).

As the antrum forms, a single layer of granulosa cells remains next to the zona pellucida. This is called the corona radiata. This too presents a fertilization barrier for sperm. Other granulosa cells (i.e., cumulus oophorus) surround the corona cells and also connect the egg to the surrounding stromal cells that have formed the outer compartment for the expanding follicle.

While several follicles are stimulated to begin their development, typically only a single follicle expands to the point of rupturing and releasing an egg into the female reproductive tract. As soon as the oocyte, zona pellucida, and corona radiata package detaches from the walls of the follicle and is free floating in the fluid, it is called a mature or Graafian follicle. While millions of eggs are in the ovaries, only approximately 500 will mature and be released (ovulated) during the reproductive life of the female. The remainder degenerate.

Female Reproductive Tract

The female reproductive tract facilitates three critical functions to ensure fertilization and continuation of the species. First, the released oocyte is transported inside of the tract where fertilization occurs. This necessitates the second important role, a pathway through which sperm must migrate. Lastly, the tract provides a safe haven where the newly created individual can develop until he or she can survive relatively independently.

Uterine Tubes

Also called the Fallopian tubes, this portion of the female reproductive tract guides the oocyte toward the uterus, allows sperm to migrate within the tube, and provides a location for fertilization to occur. The uterine tube is connected to the uterus, and extends laterally on the right and left side as the body of the tube (isthmus) expands near the ovary to form the ampulla. This expanded region is the location where successful fertilization must occur for implantation and pregnancy to occur. The opening of the tube, the infundibulum, is bounded by many fingerlike processes (fimbriae), which surround the ovary and aid in funneling the ovulated oocyte into the tract. Additionally, some of the cells that line the uterine tube have cilia on their surface that move and create a current toward the uterus, thus pulling the oocyte in the correct direction.

Uterus

The uterus is the expanded, muscular portion of the reproductive tract in which the developing individual is protected and nourished during pregnancy. Because of the thick muscular layers, this is also the engine that drives the delivery of the baby at parturition (childbirth).

Anatomy

Positioned between the right and left fallopian tubes along the midline of the pelvic region, the uterus is also located between the urinary bladder and the rectum. Initially around 50–60 grams and only about 3 inches long, the uterus can expand greatly to accommodate the growing baby during pregnancy. The top or superior part of the uterus is the fundus (much like that of the stomach in shape), the middle portion and the majority of the uterus is the body, and the narrowed neck of the inferior part of the uterus that is attached to the vagina is the cervix. Two layers make up the wall of the uterus: the inner endometrium and outer myometrium.

The inner endometrium layer facilitates implantation of the developing individual, initially nourishes the growing embryo, and assists in the formation of the placenta. Each month, this part of the uterine wall expands and becomes populated with spiraling blood vessels and endometrial glands in preparation for possible implantation of an embryo. However, if implantation does not occur within a defined amount of time (implantation window), the bulk of the endometrium (the surface functional layer) is shed during a menstrual period (menses). The base (basalis layer) of the endometrium remains and divides to regrow the thick endometrium for the next monthly cycle.

The myometrium is divided into three layers of smooth muscle that respond to hormonal signals near the end of pregnancy and spontaneously and rhythmically contract to deliver the baby. Signaled by the baby, the mother's pituitary gland secretes oxytocin that leads to the initial contractions. These contractions lead to an increase in oxytocin production, which causes stronger and more frequent contractions. This cycle continues until the baby and then the placenta are delivered, at which time the contractions slow and stop.

What are the initial precursor contractions that precede labor called?

Braxton-Hicks contractions (practice contractions) are myometrial contractions caused by nonhormonal signals and are not associated with actual labor.

Vagina

The vagina is the terminal end of the female reproductive tract and functions as the female copulatory (sex) organ, as well as providing a pathway for the delivery of the baby. The male penis is inserted into this muscular tube during intercourse. Mucus glands and other secretory cells that are essential for the last developmental stage of sperm maturation are present in the wall of the vagina. Ejaculated sperm are incapable of fertilizing an egg. Capacitation (maturation) of the sperm occurs within the vagina, and results in the sperm's ability to fertilize the egg. While the inferior end of the vagina is open, the superior end is bounded by the cervix, which contains a small opening through which the sperm must pass to get to the remainder of the tract.

External Genitalia

Made up of the same developmental building materials that shaped the male genitalia, the external genitalia of females remains open as fleshy folds of tissue that surround the opening of the vagina and urethra. Additionally, the phallic tissue that became the penile head (glans penis) in males is formed into the clitoris in females.

Labia

The two paired folds of tissue surrounding the opening of the vagina are the labia and collectively form the vulva. The outermost pair, which is thicker, fleshier, and usually covered in hair, is the labia majora. These are produced from the same tissue that was used to make the scrotum in males (labioscrotal swellings). Deeper and more medial are the labia minora. These are the thinner, more elongated folds that immediately surround the cavity which leads to the vagina and the urethra (vulval vestibule).

Clitoris

The phallic organ for females is the clitoris. The clitoris is a bundle of highly sensitive tissue located at the ventral junction of the labia (both minora and majora). It plays a major role in female arousal during intercourse and leads to orgasm and spasmodic contractions of the vaginal walls, which can assist in propelling sperm along the female reproductive tract.

Female Sex Hormones
These steroid hormones regulate both the menstrual cycle (egg production) and the estrus cycle (sexual desire) in females.

Estrogen
This is actually a group of steroid hormones that includes estradiol, which is the most abundant of the estrogens during the reproductive years of females. For women, estrogens lead to the expression of the secondary female characteristics, such as enlarged breasts, and also functions in the menstrual cycle to expand the endometrial layer.

Progesterone
Progesterone, the predominant progestogen in the female body, is also involved in similar functions as estrogens. This hormone, often called the hormone of pregnancy, affects the breasts and leads to milk production and lactation. It also plays a cyclical role along with estrogen in the menstrual cycle of females.

Reproductive Cycle
The monthly reproductive cycle of females is often called the menstrual cycle because, if fertilization and implantation does not occur, menses (shedding of the endometrium and bleeding) will proceed.

Follicle Maturation

The first half of the menstrual cycle, the follicular phase, begins with the development of several primordial follicles. This is triggered by an increase in the production of the pituitary hormone called follicle-stimulating hormone (FSH). The levels of FSH peak in the first few days of the cycle and slowly taper off until around the twelfth day when there is a spike of luteinizing hormone (LH), which triggers ovulation. At this point, estrogen has already begun rising and also peaks at ovulation before starting to decrease. Progesterone levels remain low during this phase.

While these hormones are directing follicular maturation, they also cause the endometrium to expand greatly. This is called the proliferative phase for the uterine lining.

Ovulation

As the liquor folliculi of the Graafian follicle increase beyond the capacity of the thecal cells and the ovarian capsule, the follicle and ovary rupture and the free-floating unit of oocyte and its supportive cells are expelled into the abdominal cavity. However, since the infundibulum of the uterine tube is nearby, the oocyte usually makes its way into the female reproductive tract easily. This final increase in volume and pressure is triggered by a spike in the level of LH from the pituitary gland. Also at this time, levels of progesterone begin to slowly increase, to peak in the middle of the next stage.

Luteal Phase

FSH and LH trigger the remaining thecal cells (lutein cells) of the follicular remnant to transform into a corpus luteum, a temporary structure capable of producing both progesterone and estrogen. This causes the pituitary to shut down both FSH and LH production.

However, these hormones are essential to maintaining the function of the corpus luteum. If pregnancy does not occur, it is only a matter of time before the corpus luteum destructs. Then progesterone and estrogen levels drop, which is the trigger for menses.

This phase also corresponds to the secretory period for the uterus, in which the endometrium produces endometrial glandular secretions to nourish and support an implanted embryo if this should occur.

If pregnancy does occur, the developing embryo and precursor cells of the placenta produce hormones, which keep the corpus luteum active for a much longer period of time to prevent the loss of the endometrial lining (the embryo depends on this). The major hormone produced is human chorionic gonadotropin (HCG), which is the target for many early pregnancy tests.

What is menopause?

This phase marks the end of the reproductive cycle for females. It is characterized by the absence of menses (a.k.a. amenorrhea). This is often caused by slow changes in the hormone levels in the body, which leads to the failure of ovarian follicles' ability to mature, ovulate, and shed the endometrial lining.

FEMALE REPRODUCTIVE SYSTEM DISEASES AND DISORDERS

Common Problems in Female Reproduction

Two major issues associated with female reproduction are female infertility and ectopic pregnancy. Either of these problems could lead to severe physical and psychological stresses for females, as well as create dire medical complications that could end in death.

Female Infertility

The principal cause of female infertility is a blockage of the uterine tube that prevents sperm from gaining access to the ovulated egg and/or prevents a fertilized egg from reaching the uterus. A common cause of blockage is endometriosis. In this condition, endometrial cells inadvertently gain access to the lining of the uterine tube. These cells grow and respond to the cyclical hormones, yet do not have the mechanism by which to be removed each month. This results in a continued growth into the tube and eventual blockage of the tube by this misplaced tissue.

Another condition that can lead to tube blockage from the other end of the tube is polycystic ovary syndrome (PCOS). In this condition, several follicles develop but are not ovulated and do not regress. Eventually, these cysts may rupture and eject cellular material and debris into the fallopian tube (not normally present in that amount), blocking the tube.

Uterine fibroids

Small, benign tumors called uterine fibroids are common in women, and in some cases they interfere with fertility by changing the shape of the uterus or blocking the fallopian tubes.

Ectopic Pregnancy

Even with open uterine tubes, there is no guarantee that a fertilized egg always makes it into the tube. In fact, many times these fertilized eggs remain in the abdominal cavity and attach to tissue other than inside of the uterus. This can be the outside of the uterine tube itself, in which case these are often referred to as tubal pregnancies. However, any tissue can be a target for the implanting embryo, which use enzymes to attach to anything. These are often discovered due to intense abdominal pain and may lead to removal of portions of the reproductive tract if too much damage has already been caused.

Yeast infection

While not every female will experience infertility or ectopic pregnancy, approximately 75 percent will be affected by yeast infections at some point in their lives. The yeast, *Candida albicans*, is an organism common to the female reproductive tract in small numbers. Thriving in the warm, moist environment of the vulva and vagina, something triggers the overgrowth of these organisms to cause the infection, which is characterized by vaginal itching and burning. The common method of treatment is either antifungal ointment or suppositories.

INDEX

ANATOMY 101

Epithelium, respiratory, 215–16
Epsilon cells, 237
Erectile dysfunction (ED), 266
Erythrocytes. *See* Red blood cells (RBCs)
Erythropoiesis, 82, 158
Erythropoietin (EPO), 82, 158, 171
Esophagus, 201
Estrogen, 96, 274, 275–76
Exhalation, 223
Exocytosis, 37
Extensors, 100
Extratesticular reproductive tract, 259–60
Eye, vision and, 134–36, 141

Facial muscles, 98
Facial skeleton, 85
Fats, nutrition and, 211
Feet, 91
Female reproductive system. *See* Reproductive
   system, female
Femoral vein, 155
Femur, 90, 91
Fenestrae/fenestrated capillaries, 152, 247–48
Fermentation, 51
Fever, 185
Fibrinogen, 166
Fibrous joints, 92
Fibula, 90, 91
Flexors, 100
Follicles, egg, 268, 269–70, 275
Follicles, hair, 71, 72, 74, 77
Follicles, thyroid, 234
Follicle-stimulating hormone (FSH), 233, 275
Fontanelles, 84
Fovea, of retina, 136
Frontalis, 98

Gamma cells, 237
Gastric artery, 155
Gastric rugae, 202
Gastrocnemius muscle, 102
Gastrulation, 55–56
GERD (gastroesophageal reflux disease),
   212–13
Germ cells, 256
Germ layers, 56
Gigantism, 238–39
Globulins, 166
Glomerular capillaries, 247–48
Glomerulus, 242, 243, 251
Glucagon, 202, 236
Glucose
   carbon dioxide from, 54
   diabetes and, 239–40
   energy from, 51–52, 53, 110
   formula for, 18
   low levels in blood, 236
   other sugars and, 18–19

Gluteus muscles, 101
Glycogen, 19, 110, 202, 209, 211, 236–37
Glycolysis, 51–52
Goblet cells, 205, 216
Golgi apparatus, 33–34
Gonadotrophs, 233
Granulocytes, 162–64
Granulosa cells, 269–70
Graves' disease, 239
Guanine (G), 25, 42, 43
Gustation (taste), 140
Gut-associated lymphatic tissue (GALT), 181

Hair and nails, 71–72
Hamstring muscles, 102
Hands and arms, 89, 100
H antigen, 160–61
Head and neck, muscles of, 98–99
Hearing, sense of, ear and, 137–38, 142
Heart
   assessing, with EKG, 147–48
   chambers, 143–44
   pacemaker and conduction, 146–47
   pump system, 144–45
   rate, 148–49
   regulation, 146–49
   strength of contraction, 149
   structure of, 143–45
   valves, 144–45
Heel bone (calcaneus), 91, 102
Helicase, 43, 44–45
Helix, double. *See* DNA
Hematopoiesis, 82
Hemoglobin, 159
Hemolytic disease of the newborn, 171–72
Hemostasis, 168–69
Hepatic artery, 155
Hepatic portal vein, 155
Hepatitis, 214
HIV (human immunodeficiency virus), 192–93
Hormones
   amino acid derivatives, 230–31
   defined, 230
   functions of, 230
   pituitary gland and, 232–34
   protein, 231
   sex, female, 274
   sex, male, 263–64
Humerus, 89
Humoral immune system, 164, 186–90
Hydrogen (H), importance of, 10. *See also* pH
   (acids and bases)
Hydrogen bonds, 15–16
Hyper- and hypothyroidism, 239, 240
Hypothalamus, 233–34
Hypoxia, 227

Ig(A/D/E/G/M) isotypes, 188

ABOUT THE AUTHOR

Dr. Kevin Langford is an Associate Professor in the Biology department of Stephen F. Austin State University. During the past 12 years, Dr. Langford has taught a variety of courses including human biology, comparative anatomy, human physiology, histology, and developmental biology. His research into the cellular and molecular aspects of cardiac development have won research awards from both the American Heart Association and the National Institutes of Health. He is a member of the American Association of Anatomists and American Society for Cell Biology.

Dr. Langford is the Director of the Pre Health Professions Program at Austin State, and he has advised hundreds of successful students in the course of their studies toward professional medical training. Because of his success with students and his networking with professional schools across the state of Texas and the nation, Dr. Langford is currently serving as the Chairman of the Texas Association of Advisors of the Health Professions.